'JRN

# ENVIRONMENT AND MAN
## VOLUME ONE

# Energy Resources and the Environment

ENVIRONMENT AND MAN: VOLUME ONE

*Other titles in this Series*

Volume 2    *Food, Agriculture and the Environment*

Volume 3    *Health and the Environment*

Volume 4    *Reclamation*

# ENVIRONMENT AND MAN
## VOLUME ONE

# Energy Resources and the Environment

General Editors

## John Lenihan

O.B.E., M.Sc., Ph.D., C.Eng., F.I.E.E., F.Inst.P., F.R.S.E.

*Director of the Department of Clinical Physics and Bio-Engineering, West of Scotland Health Boards, Professor of Clinical Physics, University of Glasgow, Chairman of the Scottish Technical Education Council.*

and

## William W Fletcher

B.Sc., Ph.D., F.L.S., F.I.Biol., F.R.S.E.

*Professor of Biology and Past Dean of the School of Biological Sciences, University of Strathclyde; Chairman of the Scottish Branch of the Institute of Biology; President of the Botanical Society of Edinburgh.*

Blackie
Glasgow and London

Blackie & Son Limited
Bishopbriggs
Glasgow G64 2NZ

5 Fitzhardinge Street
London W1H 0DL

*International Standard Book Numbers*

*Paperback 0 216 90076 X*
*Hardback 0 216 90077 8*

Printed in Great Britain by
Thomson Litho Ltd., East Kilbride, Scotland

# Background to Authors

## Environment and Man: Volume One

Sir SAMUEL CURRAN, D.L., M.A., Ph.D., D.Sc., LL.D., Sc.D., F.R.S., is Principal and Vice-Chancellor of the University of Strathclyde. He has served both as a member and as chairman of various Government Councils and Committees concerned with industry and technological education, and has acted as a member of the Science Research Council and, since 1973, of the Oil Development Council for Scotland. He also serves on the Boards of a number of Industrial Companies. From 1955 to 1959 he was Chief Scientist with the United Kingdom Atomic Energy Authority. His earlier work in the University of California and at the Oak Ridge National Laboratory resulted in the discovery of the scintillation detector and the high-vacuum carbon arc in a magnetic field. His inventions include the modern proportional counter and the ultrasensitive self-shielding counter for carbon dating.

EARL F. COOK, Ph.D., is Professor of Geography and Geology, and Dean of the College of Geosciences at the Texas A & M University. From 1970 to 1972 he was Chairman of the Committee on Resource Development and Environmental Compatibility of the American Institute of Professional Geologists.

ANDREW PORTEOUS, B.Sc., Ph.D., D.Eng., C.Eng., F.I.Mech.E., M.I.Chem.E., M.I.S.W.M., is Reader in Engineering Mechanics at the Open University, and Chairman of a course entitled Environmental Control and Public Health. His research interests lie in recycling, desalination, pollution control, resource conservation and the utilization of solar energy. He is the author of *Saline Water Distillation Processes* (Longman, 1974).

Sir FRANK McFADZEAN, M.A., LL.D., is a Managing Director of the Royal Dutch Shell Group of Companies and Chairman of The Shell Transport & Trading Co. Ltd. He is Chairman of British Airways and a visiting professor at the University of Strathclyde.

H. JOHN DUNSTER, A.R.C.S., B.Sc., is Assistant Director (Operations) in the National Radiological Protection Board, with responsibility for the Board's work in assessments and for its services. He is a founder member of the Society for Radiological Protection.

BRIAN BRINKWORTH, M.Sc. (Eng.), Ph.D., C.Eng., M.I.Mech.E., M.R.Ae.S., M.Inst.P., is Reader in Mechanical Engineering at University College, Cardiff, and directs the work of the Solar Energy Unit there. He is the author of *Solar Energy for Man* (Compton Press, 1972).

# Series Foreword

MAN IS A DISCOVERING ANIMAL—SCIENCE IN THE SEVENTEENTH CENTURY, scenery in the nineteenth and now the environment. In the heyday of Victorian technology—indeed until quite recently—the environment was seen as a boundless cornucopia, to be enjoyed, plundered and re-arranged for profit.

Today many thoughtful people see the environment as a limited resource, with conservation as the influence restraining consumption. Some go further, foretelling large-scale starvation and pollution unless we turn back the clock and adopt a simpler way of life.

Extreme views—whether exuberant or gloomy—are more easily propagated, but the middle way, based on reason rather than emotion, is a better guide for future action. This series of books presents an authoritative explanation and discussion of a wide range of problems related to the environment, at a level suitable for practitioners and students in science, engineering, medicine, administration and planning. For the increasing numbers of teachers and students involved in degree and diploma courses in environmental science the series should be particularly useful, and for members of the general public willing to make a modest intellectual effort, it will be found to present a thoroughly readable account of the problems underlying the interactions between man and his environment.

# Preface

ENERGY IS ONE OF MANKIND'S BASIC NEEDS. WHATEVER ITS IMMEDIATE source—coal, oil, gas, electricity, hydro-power, wind, tides, uranium or hydrogen—it all comes from the environment and cannot be exploited without some impact on the environment, usually (but not inevitably) for the worse.

In this volume, six eminent authorities bring their wisdom and experience to bear on a fascinating variety of problems. Sir Samuel Curran, a pioneer of nuclear science, reviews man's need for energy and assesses the future potential of old and new sources. He concludes that we can probably find techniques to produce all the energy that we need, but that a wiser policy would include effective measures for conservation—and a serious look at the population explosion.

Sir Frank McFadzean also takes a broad view, examining the interplay of economics, technology and politics in the oil industry. Illuminating the political follies that have recently characterized the oil debate and spelling out some simple truths of industrial finance, he suggests that more mature international leadership will be needed to overcome the current difficulties facing producers and consumers.

Dr Earl Cook draws up a large-scale balance sheet, showing where energy comes from, how it is used and how it eventually returns to the environment, usually as low-grade heat. He offers sharp comments on the wasteful practices to which modern society is addicted—though, in common with the other contributors to this volume, he emphasizes the inherent inefficiency of many energy-conversion processes in consequence of the Second Law of Thermodynamics.

Conservation is the main theme of Dr Andrew Porteous's essay in energy accountancy. Energy is used, he explains, not only in heating, lighting and transport, but in making cement, steel and every other processed material—sometimes (as in the throwaway bottle) very extravagantly. Recycling should be encouraged for its effect on energy conservation and more should be done to recover the energy available by chemical treatment of domestic refuse.

Mr Dunster examines the often-contentious issue of the environmental impact of nuclear energy. Though safety standards (and the accident record) are much better in the nuclear industry than in any other, the fear of radioactivity is very real. Mr Dunster believes that we cannot do without nuclear energy, but that we need an overall energy policy, balancing environmental factors against economic and political considerations.

Finally Dr Brinkworth reminds us that the sun is the ultimate source of almost all of our energy, but that there is considerable scope for using solar energy directly; no possible human population would need more energy than is contained in the sunlight falling on the Sahara Desert. Solar energy can be exploited without damaging the environment and could—despite many practical difficulties—be developed usefully in prosperous countries as well as in the Third World.

Mankind is accustomed to famine, drought, war and financial disaster—but, until now, shortages of energy have been only local and temporary. The message of this book is that we can have plenty of energy without damaging the environment—but only if we find a rational policy while the problem is still manageable.

# Contents

xi

CHAPTER ONE

# ENERGY AND THE NEEDS OF MAN

Sir Samuel Curran

## What is energy?

Energy is vital to all our endeavours and, indeed, to the maintenance of life itself. Primitive man knew that he had to eat to give himself the capacity to perform his various tasks, and soon he was devising crude tools which were intended to increase his effectiveness. When he invented the axe or the hammer, he was applying the force at his command more effectively, and he later used the lever to build structures which impress us to this day. Possibly the sun-worship of ancient times arose from the realization that the energy that it supplied as heat was essential to life in many ways.

Today we are much more scientifically informed. We know that the energy flowing from the sun is trapped in fossil form in a fuel like coal. We know that the energy obtained by burning this coal and using it in the form of steam power brought about the Industrial Revolution. The industrial age, as we know it, started only two hundred years ago. Scientists like Newton elucidated the laws of motion, still valid in almost all normal situations today, and certainly Newton understood the nature of force and its connection with work and energy. His ideas were ahead of the times; it is three hundred years since he established his laws, but they were of rather academic interest till machines based on steam power were made possible with the coming of coal and steam energy.

The early machines using steam for motive power were inefficient; it was Watt with his discovery of the condenser who greatly improved efficiency and led the way to the widespread application of steam-powered machines of all kinds. The capacity of man to use fossil fuels was exploited rapidly, and modern methods of manufacture were born. It must be

1

stressed that while this change was rapid, man had successfully exploited energy in horse-driven and even man-driven machines long before the birth of the Industrial Revolution. The great leap forward came when the huge reservoir of energy trapped in coal could be successfully harnessed.

It is useful to consider in a little more detail the nature of energy, because it is not readily appreciated that energy is an elusive concept. In classical mechanics, energy and work are regarded as entirely equivalent. Under ideal conditions and in the absence of friction, when a force is applied and a body moved through a distance by means of the force, the work done is the product of the force and the distance; this is the same as the kinetic energy given to the body. Energy is conserved, though it can be converted from one form to another. We have mentioned kinetic energy, which can be regarded as energy associated with the motion of a body. We also have to consider potential energy, which is perhaps adequately defined as energy due to the position of a body. A very simple illustration is to be found in the placing of a mass of some kind at a height above the ground, say on a table. When the mass is moved to the edge of the table and allowed to fall, it gains speed during the drop and strikes the ground having acquired kinetic energy or energy of motion. The potential energy that it had due to its height above the floor is converted to kinetic energy of motion. This kinetic energy is in turn converted to other forms of energy in doing work on the floor—perhaps depressing the floor, causing sound to be emitted, and generating heat by virtue of the blow. When the balance is drawn up, we find the conversion of energy is such that the total energy is conserved. Energy is always conserved, though it may take many forms such as potential, kinetic, heat and sound.

During the last century many distinguished scientists studied radiation phenomena, especially in the form of light and heat. Towards the end of the century, in spite of successes achieved by Kelvin, Clausius and others, there were increasing difficulties with theory due to the dual nature of radiation. Some phenomena were explicable when radiation was thought to be wave-like in nature, while other phenomena like the photoelectric effect seemed to depend on radiation being particulate or, as we say today, in the nature of packets of energy or quanta of energy. Out of this problem came the ideas of quantum physics. We realize now that there is no need to exclude either interpretation, and that radiation is made up of quanta but can often be regarded as consisting of electromagnetic waves for the purpose of explaining some aspects of its behaviour. We can go further and prove that a single particle of matter may have wave-like properties. This duality of matter is accepted at the present time by scientists, and indeed recognized as being an inherent fundamental property of matter.

## Matter and energy

Most of us have preconceived notions about matter. We think we know what a mass is. We tend to regard it as a lump of some stuff or substance. In one sense this is a perfectly sound view to take of matter or mass, although there is apt to be confusion about mass and weight. To the physicist, weight is, in fact, a property of matter. It is the force the mass exerts due to the existence of the gravitational field of the earth. Since space travel has become almost commonplace, we now appreciate that objects and man himself can be weightless, though obviously still having mass or matter. So weight is a very special attribute of matter. In thinking of weight, the gravitational field between the large earth and a body is predominantly that of the earth. We should note, however, that gravity of a body exists in its own right, and that a gravitational attraction is exerted between any two masses—though it may be so small as to be negligible, except in the case of the large earth and another body.

It was Einstein who made the great break-through in our understanding of the nature of mass or matter. He proved to the satisfaction of scientists that mass was literally a form of energy. We have already talked of different forms of energy of which we are all aware. It is an immense step forward in our knowledge of nature to accept that Einstein was right in saying that mass and energy are really the same. If mass is made to disappear, an equivalent amount of energy will appear. In most phenomena with which we are familiar, at least those around us in everyday life, mass does not disappear in appreciable quantity, and in classical physics (non-relativistic physics) it was customary to regard mass as being conserved in all changes. We now know that in relativistic physics in which phenomena occur between masses which have high speeds, mass as such can disappear, and an equivalent amount of energy, often in the form of radiation, will be created. If in such phenomena we regard mass as simply a special form of energy, we find that the fundamental law stating that energy is always conserved is entirely true. Einstein expressed his law in simple mathematical form: if $E$ is the energy and $m$ the mass, then with $c$ representing the velocity of light

$$E = mc^2$$

The velocity of light in free space is a constant, and its value is given by $c = 3 \times 10^{10}$ cm/sec. We can in fact talk of the annihilation of matter, now meaning that matter in the form of mass as we know it can literally be caused to disappear and an equivalent amount of radiation (in the form of radio waves, heat rays, light, X-rays, $\gamma$-rays, etc.) emitted. The atom bomb and the hydrogen bomb are the best-known examples of man achieving the transmutation of matter to radiation. Thus kilogram quantities of an element such as plutonium can be transformed in appropriate conditions to an enormous flux of radiation. The stars burn matter in this way, and in our own star, the sun, enormous masses of hydrogen disappear in the nuclear reactions, and tremendous amounts of solar radiation escape into space, a small but very important fraction of this falling as heat and light on our own earth. We would, of course, perish if for any reason the solar furnace were quenched. Nuclear reactions in the sun are all important to our survival.

### The needs

If we look back at the history of man we realize that man is unique in his knowledge of the importance of energy, first as it affected his survival, and later as it affected crucially his standard of living. He soon learned to use tools, a facility in which man is almost alone in the animal kingdom—in a sense tools allow for effective concentration and control of energy. A hammer focuses energy well, usually for peaceful purposes, and in another context weapons such as the bow and arrow focus human energy. Indeed it was the power to utilize energy, his own energy, in an extremely versatile way that gave man in at least some directions his marked superiority over other animals. Later he came to harness the energy of other animals, and in particular the horse, which he used to perform some of the heavy tasks that faced him in, for instance, farming and building. It is still customary to measure rate of doing work or rates of expending energy in terms of horse-power (hp), which as a unit is applied in mechanical engineering, e.g. in defining the power or capacity for work of a motor-car.

It is necessary to have useful units of energy and we have many in practice. Thus scientists often refer to a foot-pound-weight, meaning the work or energy used in lifting a mass of one pound through a height of one foot against gravity, which of course gives the mass its weight. We have rates of working and the horse-power is equal to a rate of 33,000 ft-lb-weight per minute. There are corresponding units in the metric system, and likewise in different branches of science there are various units of energy, and various rates of expenditure of energy or rates of working. In electrical science we have the kilowatt as a rate of expenditure of electrical energy and the kilowatt-hour as a useful unit of energy (often referred to as a Unit). In fact the kilowatt is not so different from the horse-power. One horse-power is roughly $\frac{3}{4}$ of one kilowatt (kW). The conversion of one form of energy to another makes the relationship of the various types of unit clear. Scientists such as Joule showed that mechanical energy could be converted to heat, and we know that heat can be used to develop electrical energy.

In industry today man has turned his knowledge of energy and its control to great advantage. His needs are many and he has found a great range of ways of meeting them. One of his basic needs is food. When he turned to agriculture in order to satisfy many of his food requirements, he learned readily that heat, in the form of the sun's radiation, was vitally important. Crops grew more abundantly in warmer climates. Thus he came to depend on the sun, and in fact turned to worship the sun in many cases. Later he took to protecting his food plants from the cold, and now in many parts of the world he grows plants under glass, trapping the sun heat and reinforcing its effect by other forms of heat. So for food, man depends critically on energy, and some of this energy he obtains from fuels by the burning of wood, coal, petroleum, though in practice he may convert those fuels to electricity in power plants remote from the place of use. Likewise he uses energy to heat his own home, and within a very few centuries his ways of doing so have changed tremendously. For many thousands of years he had to use wood, in some countries peat, and in some countries to this day man uses millions of tons of animal excrement. The discovery of huge coal deposits brought about marked changes and in many countries coal is used for home heating. However, in more recent times the problems of pollution and the consequent introduction of clean-air legislation have led to a rapid move from direct to indirect use of coal. At the same time, and largely within the last hundred years, petroleum has come to replace coal in many applications. At the present day we have the large power plants operating with various fuels, of which the principal three are petroleum, coal and gas.

Thus we have examined some of the main needs of man—his need, for example, for power to drive machines in industry, particularly in factories producing many of the goods he requires. He needs energy to heat his home. He needs energy to help him produce food. He needs energy to build. One of his needs that has become of major importance is that of transport. The modern age has seen a great rate of increase in the use of transport of all kinds, particularly in the use of motor-cars by so many people in the advanced countries. The consumption of petroleum in cars has made people more mobile than at any time in the past, but it represents a large outlay in the energy expenditure of man. When petroleum supplies are interrupted for any reason, there is immediate anxiety, since cars and public-transport vehicles depend almost wholly on such supplies. The indirect use of petroleum is also very substantial. Thus electric trains depend on power from electricity generating stations, and a good fraction of those use oil as fuel. This is a fact that is sometimes not fully appreciated. If petroleum is burned to produce superheated steam to drive turbines which in turn drive generators—a usual arrangement in a conventional generating station—the shortage of petroleum means the plant has to be converted for the burning of gas (natural or synthetic) or coal. This is not practicable in the short term, so that modern transport is very vulnerable to the maintenance of the supply of petroleum.

## The sources

We have dealt in somewhat broad terms with energy and the needs for energy. It is desirable to look in more detail at the total range of energy sources available to man. We have seen that his needs are substantial and that for a few decades his consumption of energy has been rising at a high rate. In general, and certainly in the advanced countries such as the United States, consumption has been increasing at a compound rate of about 5% per annum, and this rate implies a need to provide roughly twice as much energy in a period of around 15 years. Since the United States uses about one third of the total amount of energy consumed, we can see that the advanced countries present a major problem in respect of the continued provision of very large amounts of energy. The long-term implications of such a level of consumption could be very serious indeed, and the interruption of even one source of supply could bring about extremely grave repercussions. Let us look at the principal energy sources available and their abundance. We shall begin with the most obvious and assume that conversion to electrical energy is feasible if this is desirable or necessary.

*Hydroelectric power*
The potential energy of a body has been mentioned, and the potential energy of large masses of water is an important source of power. The sun heats the waters of the oceans, seas and lakes, and causes evaporation of some of the water. This water may form clouds, and part of the water in these clouds may be precipitated over comparatively high land. Such water, trapped in lakes or man-made reservoirs by means of dams, can be released as required, and the kinetic energy derived from the fall of the water to lower levels may be used in driving turbines, which in turn drive generators to yield electrical energy. We seem to have here a clean source of power of unlimited duration, but the real facts of the situation belie this assumption. The countries which have comparatively large rainfall and land mass of comparatively high elevation are rather few, and not many countries are able to produce a substantial fraction of their electrical energy by hydroelectric means. At the same time there is some limitation due to the inefficiency of the energy conversion. We shall have more to say later about the rather low efficiency obtained in most systems which provide energy in electrical form.

We note here, however, that when we burn fuels such as oil or gas to produce high-pressure steam, which in turn is able to drive turbines coupled to electrical generators, the overall efficiency is usually less than 40%. In the direct combustion of the fuel to give heat energy, as in heating a building or home, the efficiency can be higher, and the whole question of efficiency of use of energy is at the heart of many of our problems.

The original hydropower was not electrical in form. Water flow was used to rotate mill wheels and in other ways put to direct use. The introduction of the stage of conversion to electrical energy resulted in a decrease of efficiency, but the price is paid for the sake of the major attraction of easy transmission to the user. Electrical transmission is a large subject in its own right and, at the high voltages normally in use today, usually ranging from around 30,000 volts to 300,000 volts, it is relatively loss-free.

*Thermal energy of the earth*
Some natural sources of power are as yet comparatively modest in their contribution to the total energy needs of the modern world. The earth is a planet of the sun, and we know that there is hot molten material within the crust. In fact the crust sometimes develops faults, giving the possibility of tapping thermal energy. While such energy may be extremely valuable in a few countries (parts of Italy, Iceland and New Zealand, for instance) the steam has to be utilized near to the source itself, or the system is comparatively inefficient. It is possible that with considerable research effort, in the future the vast amount of thermal energy stored within the earth could be tapped with great benefit. The low thermal conductivity of the crust of the earth makes the process an extremely difficult one, but work is beginning.

*Wind and tides*
The great power in the movement of massive amounts of air as winds is attractive, but to exploit wind power is by no means simple. The energy in the great air masses is enormous, and it is the problem of utilizing part of this energy that has been found intractable to date. We know that in the Netherlands and in a few other countries windmills proved of considerable value, but among other difficulties is that of the variability of the wind. One other problem is the size of the structure required. Windmills able to produce around 100 kW have been built and tested, but the structures proved too fragile to withstand the buffeting of strong winds. In spite of such problems it is likely that in parts of the world where strong winds are frequently experienced some revival of interest will be seen. As structural materials are improved and energy storage becomes more practical, the windmill could prove valuable. As we have noted, the

uncertainty of wind makes energy production somewhat unpredictable, but this difficulty could be largely overcome if the problem of storage of energy was solved.

One method of storage now available and in which improvements can be expected is that of hydrogen production. If we use the windmill to drive an electrical generator, we can, when the power is not required by customers, use it to electrolyse water, and the hydrogen so released may be stored as a fuel. This hydrogen fuel can again be used directly, or indirectly to produce electricity. In this way the energy supply could be continuous, though we should note that the various conversion stages mentioned are somewhat wasteful, and the overall efficiency could not be high. On the other hand, the primary energy source, the winds, do not cost anything.

The opportunities for further research in geothermal energy and wind power are considerable, though it must be admitted that the amounts of energy available, even if the researches are successful, are not large in relation to the total needs.

Tidal energy has been the subject of a good deal of research and it will no doubt continue to attract attention. It is clean and reliable but so far has not been impressive on a world energy basis and, except in special circumstances, is unlikely to prove of great value when set alongside the other sources of supply. More recently, the possibility of harnessing the energy in sea waves has been examined, and this may in the longer term prove to be an attractive and useful programme.

### Fossil fuels

Without doubt the truly large-scale sources available to man at the present time are the fossil fuels. Of these the most abundant is coal. Even at the present-day level of consumption it is estimated that coal reserves that can be won with near certainty will meet estimated needs for many hundreds of years. In considering coal, we can allow for a rather substantial rate of increase in demand, say, doubling in 20 years, and still meet all major requirements for several hundred years.

The position regarding oil and gas is not nearly so reassuring. Some experts consider that if the present level and rate of increase of use of oil continues, there could be serious shortage in as little as thirty years. However, the reserves of oil and natural gas are by no means completely known, and most forecasts are based on rather tentative estimates of the likelihood of finding new fields, especially off-shore fields. Such fields continue to be found, but the rate of increase of known reserves is less than the rate of increase of consumption. For the present we can state confidently that coal and oil will provide the most abundant source of energy throughout the twentieth century.

The importance of oil is examined in detail in chapter four.


## Scale of use of energy

The appetite of man for energy is enormous. We have underlined the fact that man has for ages tried to liberate himself from dependence on his own manpower in manufacturing, in building, and in agriculture. The last hundred years have seen a staggering increase in his capacity in this respect, but it must be noted that this has been achieved at the cost of rapid depletion of most well-established energy supplies. Considering man as a machine, we know that on average he needs to be supplied with around 3000 large calories of energy, in the form of food, each day. But he is not efficient as a machine, being in practice able to provide an energy output as work of only about 10% of the input, around 300 large calories. We can in fact say that as a machine he is roughly equivalent to an electrical motor of about $\frac{1}{5}$ to $\frac{1}{7}$ hp. Yet the power consumption in a country like the United States corresponds to approximately 100 times this amount per person. In other words, each inhabitant of a country like the United States has in a sense some 100 slave-equivalents working for him or her. This is the order of magnitude of the energy utilization in a number of advanced countries, though the United States, Sweden, the United Kingdom and certain European countries are particularly heavy consumers of energy.

Where does all the energy go? The manufacture of iron and steel in large quantities in any industrial country implies an unusually big energy consumption as estimated on a per capita basis. The fact that the per capita energy consumption is high does not necessarily imply wasteful or lavish expenditure of energy. The processing of materials like iron ore in vast amounts means inevitably that the energy required to decompose the iron oxides is consumed. Most metal production processes require considerable amounts of energy, sometimes much in excess of those required in the industries using the products of iron, though fabrication itself can be expensive in energy.

In stressing this requirement of the heavy industries such as metal production, it is wise to point out that agriculture uses large amounts of energy. Much of agriculture today involves the use of chemical fertilizers and pesticides. Recent studies by experts concerned with energy policy have revealed some very interesting aspects of this subject. Chemical fertilizers are energy-expensive in production, and it has been pointed out that the cost in energy terms of materials used in cultivating crops for human and animal consumption may be high. In some cases the energy yield of the foodstuff may be less than the energy expended in cultivating the required crops. Intensive farming has to be reviewed in the light of this somewhat surprising fact. It is true that the per capita consumption in an agricultural community may appear to be low, but the use of energy in manufacturing fertilizers and other necessities produced elsewhere and imported into the agricultural area may make the energy balance sheet look very different when the analysis is done in full detail.

*Solar energy*
It is logical to proceed now to some preliminary discussion of solar energy. (A more detailed appraisal is presented in chapter six.) In the final analysis most of the energy sources known to us depend on solar energy. We have given several instances of the validity of this point of view. Solar radiation causes evaporation of water from the seas and surface of the earth, and hence is the basis of hydropower; sunlight is the basic source of the energy of plants, and thus of all living organisms which give rise to the energy stored in coal and petroleum; solar radiation is involved in generation of winds; and many other illustrations can be produced. However, there is possibly in geothermal energy an exception to the rule. Much of the energy stored in the earth within the crust is due to the disintegration of radioactive substances such as uranium, thorium, rubidium and potassium, which occur naturally. We shall have more to say about nuclear power, which also involves radioactivity.

Meanwhile it must be stressed that in terms of available energy, the amount of sunlight falling on the earth exceeds our present total energy consumption by a very large factor, greatly in excess of 1000. If we could efficiently harness as much as $\frac{1}{1000}$ of the solar radiation, we would be able to meet all our needs very readily. It is essential that this possibility should be explored, and much thought and effort is presently directed to this end.

Solar cells of various kinds have been produced, including silicon cells and cadmium sulphide cells. When light falls on such cells, a photo-voltaic effect is observed, and current can be drawn from the cells. The voltage generated and the level of current that can be generated in this way are small, typically between $\frac{1}{2}$ volt and 1 volt at a current of about one thousandth of an ampere. Thus a single cell is capable of producing around one milliwatt of electrical power (one watt is equal to one volt multiplied by one ampere). The cells can be very small indeed, and arrays of such cells, mounted for instance on thin sheets of suitable plastic, can be connected in series and parallel so that a power source of usable voltage and current capacity can be made. Thus, in principle, energy sources for domestic and industrial applications are available now. There are difficulties nevertheless. The efficiency of conversion of the solar energy to electrical energy is usually found to be small, perhaps in the range 2% to 5%; there have recently been claims for a much higher efficiency of around 18%. The area of the solar convertor has to be large, and in turn this can mean a costly capital investment for a really useful source. At present the practical applications are not numerous, being confined to use in space vehicles and other somewhat exotic equipments. The future of the method must be seen as very promising since (at least in principle) it assures us of considerable energy availability for the foreseeable future, even if the installation costs continue to be high. The method is apparently free of pollution problems, a matter of increasing relevance. Storage of the electrical energy as hydrogen would remove dependence on the constancy of availability of sufficient radiation. At present, however, a large array would probably yield less than one kilowatt of power.

An intriguing version of the use of sunshine comes from the biological field. Efforts to grow crops which are efficient in storing solar energy are increasing. Such crops can be treated to yield organic substances, which in turn could replace the fossil fuels in many applications. Again overall efficiency of conversion of the energy is low, at best 1%, and this may prevent the production of useful energy in commercially valuable amounts. Furthermore we have noted earlier that the manufacture of fertilizers, pesticides and other essential substances for intensive use of agricultural land itself involves the expenditure of substantial energy in the chemical production processes. The overall return, regarded strictly from the energy angle, may be severely limited.

*Nuclear energy*

We have stressed the major importance of the solar radiation in supplying directly and indirectly so much of the total energy needs of man. The sun is the most important of all of our energy sources, and it is found to be a nuclear furnace of enormous output. An immense tonnage of hydrogen is consumed in fusion every day.

In February 1939 the real explanation of the phenomenon of nuclear fission was put forward. Work on this remarkable process proceeded rapidly but with increasing secrecy. It was realized that the process was novel in several respects; for instance, it gave promise of chain reactions similar in principle to those observed with combustible chemical substances, and hence also possibly able to yield energy at an explosive rate in the same way as highly explosive chemicals. There was, however, another reason for security and secrecy. If nuclear reactions could be self-maintaining and possibly explosive in their burning, the yield per unit mass of fuel would be around one million times greater than with typical chemically reacting substances.

In principle it is easy to explain the process of fission. Uranium is the heaviest of the naturally occurring relatively abundant elements. It contains within its nucleus a substantial preponderance of neutrons as compared with protons. Neutrons and protons in close association compose all known nuclei. In the heavier elements the number of neutrons exceeds the number of protons to the point where the element does not have a completely stable nucleus; uranium is an example. Many heavy nuclei are unstable and transmute with the passage of time. Uranium is radioactive—as are various other naturally occurring elements—but the uranium nucleus can also change by the unusual process of fission, or splitting into two lighter nuclei. In doing so it releases at the same instant a small number of neutrons, generally two or three in each act of fission.

Now the remarkable nature of the experiments of 1939 can be understood. It was realized that a single neutron could bring about the fission of a uranium nucleus and that at the same time more than a single neutron would be made available in the same act of fission. The extra neutron entering the uranium nucleus was, so to speak, the straw breaking the camel's back. A suitable mass of uranium could in principle burn or perhaps explode with the release of a vast store of nuclear energy. Each act of fission of a single atom yields approximately 200 million electron-volts (using the typical unit of energy for nuclear reactions) and this is to be compared with the corresponding yield in chemical reactions of a few electron-volts per reacting atom or molecule. Thus, in theory the energy yield per unit mass could be approximately 1 to 10 million times greater when nuclei undergo fission as compared with atoms undergoing exothermic chemical reactions. It was this kind of reasoning that underlay the secrecy that soon surrounded work on the atom bomb (which could more accurately be called the nuclear bomb).

The bomb did not come first. By 1942 in the Argonne Laboratory in Chicago the first nuclear reactor was operating. In this type of power plant the neutrons were moderated by means of pure graphite in which the uranium fuel rods were imbedded. Moderation or slowing-up of the neutrons depended on the collisions of the neutrons with the light nuclei of carbon in graphite form, and the slowing-up allowed steady operation and control of the level of output of the reactor. Essentially the reactor is simply a source of heat in which the fission within the uranium rods results in the generation of heat, mainly by the dissipation of the high kinetic energy of the fission fragments. Other heating processes are involved, like that arising from the formation and subsequent decay of radioactive elements (the fission products themselves).

The nuclear-energy industry is discussed in chapter five.

### Bombs and power plants

In 1942 the introduction of a large new source of energy by the nuclear fission of uranium, and possibly of thorium, could be foreseen. In 1945 it was shown to be practicable to generate the power explosively. Both uranium (in the form of the lighter isotope of the element, i.e. uranium 235 as opposed to the much more abundant isotope uranium 238) and plutonium were used in the construction of atomic bombs. The plutonium itself was the byproduct of the fission of uranium in nuclear reactors. When uranium 238, which represents 99·3% of the natural element, absorbs a neutron (this is simply arranged in any reactor) the resulting element (uranium 239) soon changes to plutonium 239. This element, like uranium 235 burns, as it were, in any suitably arranged mass of the element above the so-called critical amount. It is thus a short step to envisaging the possibility of using uranium 235 and (even more important in the longer term) uranium 238 as nuclear fuels. If the reactor is designed with sufficient care, the consumption of uranium can be regarded as providing energy when power-production proceeds, and as simultaneously yielding a new fuel (plutonium) in amounts greater than the amount of uranium consumed. This breeding process depends rather critically on the control of the neutron balance in the reactor, but it is possible to produce useful amounts of power and to convert a good deal of the uranium 238 into plutonium for later use. This process runs on fast neutrons, with no need for a moderator—hence the name *fast reactor* is applied.

A considerable amount of electrical power for industrial and commercial use is now generated in nuclear reactors. In the United States, some thirty years after the war-time studies, nearly two hundred power reactors are in operation, and each of the more modern type generates around 500 MW (megawatts). This is a very significant contribution to the power needs of the country with the most substantial per capita energy consumption. Only 6% of the population of the world resides in the United States, but the country uses almost 30% of the total power consumed on earth.

Many questions are posed by the introduction on a large scale of nuclear energy. Experts tend to disagree about the amount of uranium that is available. At the time the programme started, in the late 1940s, there was high hope, at least on the part of the scientists, that in uranium man had found a very desirable form of energy, available for hundreds if not thousands of years, and in amounts sufficient to satisfy the majority of man's energy needs. In more recent times grave doubts have been expressed as to the validity of this optimistic view. Is there real justification for such misgivings and fears?

Geologists differ in their estimates of the amount of useful uranium that could be exploited commercially. It is probably fair to assume that usable deposits can be obtained readily to supply total present world energy needs for some hundreds of years. It is much less clear what is the total available ore supply, so for the present it would be unwise to assume that a supply capable of lasting more than a few hundred years is available. If we assume successful breeding in a fast reactor (the only form of reactor that breeds effectively) then the duration of the uranium age in power could be increased very considerably. With uranium and plutonium efficiently used in breeder power reactors, most of the total world needs could certainly be met for a good many hundreds of years.

There are objections in principle to breeding. The main fuel for power generation would increasingly be plutonium, and plutonium is a potentially very obnoxious element. Small amounts, even perhaps a single microscopic particle inhaled into the lungs, could conceivably prove to be carcinogenic. Thus the widespread use of plutonium depends on very high standards of safety in the operation of the reactor. The fast breeder reactor is, however, less readily designed for complete safety than the slow moderated power reactors presently employed as electrical power generators.

In both types of reactor there is the problem of the disposal of radioactive waste products. Many reactor installations retain such products in large storage tanks, often in liquid form, but some experts do not consider this a satisfactory long-term answer. It is argued that leaks could occur and possibly contaminate water supplies.

## Prospects and pitfalls

It is interesting to speculate on the probable outcome of further research. It is conceivable that concentrated waste could be frozen deep in the ice at the Poles, or perhaps buried in a large natural salt bed; possibly it could be shot into space or perhaps at sufficient concentration it could make its container hot enough to melt its way into the depths of the earth's crust. However progress comes, it is conceivable that a foolproof method could be found. Here a further argument is advanced. In an age when hijacking is commonplace, it is contended that considerable hazards to human life would be presented by those who stole large amounts of radioactive material and used the material to threaten human life. The most dangerous material in many ways would be plutonium, capable of endangering life on account of its radioactivity, and capable moreover of being made into a small yet powerful nuclear explosive device.

It is sad to reflect that some of the early hopes of the nuclear physicists who worked on the reactor problems were based on the assumption that coal-mining was an undesirable dirty job and winning sufficient uranium ore would be a much cleaner simpler task. They also held the view that the smoke, dust and dirt associated with coal burning would be removed, and conditions in industrial life and conditions in our cities greatly improved. In the past two or three decades, however, the burning of coal in large generating stations has been vastly improved technically, and at the same time the conditions under which coal is mined have been greatly improved. Nevertheless coal is by no means an ideal fuel and, even when much of the dirt and disease associated with it has been eliminated, the immense amounts of carbon dioxide produced and passed into the atmosphere could have adverse effects.

The environmental problems involved in the use of fuels of all kinds receive today a great deal of attention and study. This does not imply that they did not affect the attitude of scientists and engineers in past decades. As already mentioned, one source of inspiration of the early workers in atomic energy or nuclear power for peaceful purposes was the possibility of liberating man from the necessity of winning coal in the uncomfortable and hazardous conditions in many coal mines. Today, however, the care of the environment features more prominently than ever before. Partly this increased concern stems from the realization that pollution of the atmosphere can adversely affect human health and well-being.

One obvious example of this is often quoted. The use of automobiles in large numbers in the city of Los Angeles gives rise to a high concentration of exhaust gases, and these gases and vapours are affected by sunlight so that an unhealthy smog frequently results. Steps are now being taken to reduce the quantity of undesirable gases and vapours exhausted into the atmosphere, and recent research looks extremely promising. Other examples of the importance attached to care of the environment can be instanced. In New York the use of oil-burning appliances is widespread, both in large generating stations and in buildings of all kinds, and in homes. The sulphur dioxide that is a product of this burning (and which has its origin in the small percentage of sulphur in the petroleum) is particularly harmful in many ways. Life can be made uncomfortable for everyone, and in fact there is a substantial hazard to health in the worst conditions. In consequence, only petroleum with a small sulphur contamination is imported into the city. For similar reasons the city of Tokyo, and the industrial belt to the south of the city, are rapidly converting to the use of imported natural gas which is comparatively free of sulphur and which therefore burns with less deleterious effects on the atmosphere.

### The cost of energy

We have seen that there is a wide range of possible energy sources available, but we have indicated that many of these energy sources present problems in their use. Some of the problems are those of availability, which must feature prominently in any estimate of the probable return on a large capital investment. This is of particular importance when considering the construction of a considerable number of major generating plants. The question of hazard and the effect on the environment are likewise involved in decision-making in the power field. In addition, the cost of a unit of energy is of critical importance. Here again we enter a rather speculative field.

Oil is a major energy source and its transportability is a valuable feature in its favour. It can be used in the home as well as in central plants. It is a versatile energy source, much more so than nuclear fuel, which at present must be used to generate electrical energy, as only in this form can it be transmitted to the individual customers. This transport ability of oil has in practice made personalized transport available. The internal-combustion engine cannot be powered by nuclear fuel and only if the electrically-driven car becomes practicable will nuclear power compete in this field. This need to convert to electrical energy affects the price of the energy supplied to the customer.

It is necessary to stress that the cost of a unit of energy supplied from different sources is still subject to many uncertainties. Oil is readily obtained from many oil fields, but the geographical distribution of the fields is in no way uniform. Recent explorations suggest that substantial oil deposits are more widely spread than was believed even ten years ago. The fact remains that it is often not an expensive fuel to obtain as raw material and the price is in some respects artificial. If most of the major sources are in a few countries, and there is any shortage of supply, then the price charged may be artificially boosted. While energy was abundantly available throughout the industrialized world, the oil obtained from the Middle East was relatively inexpensive, but now the price is increased to a level where nuclear power is well able to offer cost benefits. Coal mining is a rather labour-intensive industry and, if wages are high, then the price of coal-derived energy is bound to be high.

In considering costs, therefore, we have to consider a number of factors which themselves involve many uncertainties. Energy is a basic need in industry and, while in the past it could be regarded as a relatively small component in the cost of most manufactured products and materials, this is rapidly becoming less and less the case. If for any reason the energy supply falls far short of total consumption, the consequences so far as our standards of living are concerned can be extremely unpleasant.

The needs of man change from age to age, but it can be maintained that the change is always upward. Many things that would have been regarded as luxuries a few decades ago are now considered to be necessities by almost everyone. Perhaps we can illustrate by mentioning electrical lighting and heating in our homes. These are recent innovations, but how many would tolerate candlelight nowadays? We have also radio, television, telephones, not to mention central heating in homes. The range of foodstuffs, the variety of clothing, the diversity of amusements, the vacations (often by air transport), the car, all represent the new necessities of life. In building we have new materials in bewildering array. Thus aluminium is used in great amounts, of the order of millions of tons per annum in a typical advanced country. Immense amounts of wood are still used, particularly in satisfying the demand for newsprint.

Most of these products can only be made available by the expenditure of immense amounts of energy. Consider aluminium. Plants for its production often depend on the availability of considerable hydroelectric power, as the electrolytic separation depends on the passage of large currents. If the energy supply available in any of the advanced countries falls by even 5% there is a difficult situation, and 10% is critical in some particular areas. Meanwhile most of the advanced countries have tended to use 5% more energy per annum. Clearly there is a limit to how long such a rate of increase can be sustained, and some experts believe the advanced countries must begin to conserve energy and even decrease consumption.

The problem of more economic use of energy sources offers a real challenge to scientists and engineers. We have pointed out that conversion of fuels to electrical energy (fissionable material in a nuclear reactor is such a fuel) gives rise to a serious loss of energy, and the overall efficiency of a plant is generally around 40%. This means that 60% of the energy of the fuel is lost. Such a huge loss is a matter of great concern. There are hopes that a good part of this energy loss can be avoided. In the nuclear power field we find that effort is turned towards the high-temperature gas-cooled reactor, partly because high temperature in the core gives a real possibility of increased efficiency. Work on the separation of hydrogen from water at such high temperatures (around 800 to 900°C) indicates that by effecting the separation through chemical chain reactions, the overall efficiency might be raised to around 70%. This would represent a vast improvement on existing practice, and perhaps it is along such lines that the increased need for energy should be met. The energy in practice would be as hydrogen which would be well able in principle to satisfy all the diverse energy needs that arise in an advanced industrialized country.

It has been proposed that huge nuclear reactors giving about 100 times the energy of the largest so far constructed should be built, and the hydrogen produced in these great reactor plants distributed over considerable distances—hundreds of miles if necessary. In principle, if we can raise efficiency and find uranium in amounts that most geologists regard as reasonable, a nuclear power age lasting at least hundreds of years is quite possible. We have said already that coal reserves are such that the same claim can be made for coal. For oil there is much less ground for optimism, and thirty to one hundred years is the kind of range visualized.

## Carbon dioxide in the atmosphere

All three main established energy sources, coal, oil and nuclear, have inbuilt difficulties. We have discussed the difficulties of the nuclear fission answer, and for oil and coal we have touched on the problem of the production of large amounts of carbon dioxide, and the possibility that such amounts continually vented into the atmosphere might present us with real problems. We did not stress this difficulty to any great extent, but there are some experts who consider that carbon dioxide release might pose a real threat. The part played by the normal small percentage of carbon dioxide in the atmosphere (0·033%) is critically important, acting as a kind of blanket for the earth and retaining heat, particularly the solar heat transmitted to the earth's surface.

It is argued that the amount added to our atmosphere by the burning of fuels could alter the amount of carbon dioxide to an extent which might modify the blanket action, with serious effects on the weather. On the other hand, some scientists believe the processes are not so critical, and that mixing with the oceans might stabilize the carbon dioxide. There is thus disagreement in a matter which could prove crucially important—the nature of weather on earth. In such circumstances it can be maintained that rapid change of carbon dioxide production should not be permitted. At the present rate of release of the gas into the atmosphere experience of a few more decades should help to resolve the question. If the carbon dioxide problem is found to be less serious than some theorists believe, then more latitude in passing the gas into the atmosphere can be allowed.

We note in passing there is nothing very new in the release by man of the gas in considerable amounts. Forest fires have made huge quantities and, in countries such as India, the burning of some hundreds of millions of tons per annum of refuse, dung and other materials has been practised for hundreds of years. However, the modern industrial age has speeded up the process by a large factor, which means we are not at all certain of the final outcome of the release of such huge amounts of carbon dioxide. In considering the three major fuels, we have in the carbon dioxide question a point in favour of nuclear power.

Environmental considerations may prove to be most important of all in determining the energy policy of the future. We have discussed possible weather changes, but the carbon dioxide of the atmosphere is involved in biological processes of all kinds. We are familiar with the part played by carbon dioxide in plant life, and here again we cannot be sure what result might follow from a large increase in the amount of carbon dioxide in the atmosphere.

### The long-term problem

We have explored the advantages and disadvantages of the principal energy sources available to us, and established that there is no easy answer to the question of energy policy for the future. The sources that would seem to be near to the ideal, e.g. solar cells, are not yet practical so far as the major needs are concerned. We come then to examine the feasibility of a project which could in some ways offer great advantages. We refer to the process of nuclear fusion.

We stressed the fact that most of our energy sources owe their origin to the radiation of the sun. This is clearly true of coal and oil. The mode of operation of the sun itself as a powerful source of radiation is worthy of study. The sun is a typical hot star, and much of what applies to the sun applies equally well to countless other hot stars. It is possible that, through knowledge of the processes of energy release in the sun, we can reproduce the same reactions on earth and come to have a new type of power source within our control.

The energy of the sun is of the nuclear variety. It can be shown that this must be the case simply by estimating the tremendous magnitude of the flux of radiation emitted, and the calculation of the temperature that must be maintained in the interior. We know that at the surface the temperature is about 5000°C and in the centre the temperature is several million degrees centigrade. At such temperatures and with the type of pressure that exists near the centre we are in the region of nuclear reactions. Clearly any man-made machine that operates in something like the manner of the sun has to work at temperatures in the million-degree range and also at reasonably high pressures.

## Fusion energy

The sun is known to be composed mainly of hydrogen; we therefore assume that the nuclear burning of hydrogen is the most likely process that takes place within it. This is the most elementary of the fusion processes known to nuclear physicists. The nucleus of the hydrogen atom is the proton, one of the two elementary building bricks of all nuclei. In fusion we could have two protons come together to give a helium nucleus, but the helium nucleus also needs one or two neutrons in its known isotopic forms. It is possible to show that, when hydrogen is maintained at high temperature and sufficient pressure, fusion does occur and helium is formed from the hydrogen. Theoretically we can arrive at this result in several ways, but perhaps the most straightforward method is to assume that two of the nuclei of heavy hydrogen, known as deuterium and denoted by $H_1^2$, (2 is the mass and 1 is the nuclear charge) fuse together to yield a helium nucleus:

$$H_1^2 + H_1^2 \rightarrow He_2^4$$

In this process a very considerable output of energy is obtained. It is therefore theoretically possible to arrange matters suitably in a fusion reactor to bring this reaction into operation, and in certain conditions it can be shown to be self-sustaining. This means that the energy released not only gives the reacting particles sufficient energy to sustain the reaction or burning, but at the same time yields excess energy to be used in other ways.

The conditions for successfully maintaining the fusion reactions have been established in the hydrogen bomb, but in it the energy release is of very great amount and of very short duration. A typical hydrogen bomb could have a yield of the order of 1 megaton equivalent of TNT, but it lasts for a time of the order of one microsecond. This is of little value in industry, where the continuous release of energy in a controlled fashion is what is really required. So far this goal has not been attained.

It is a goal well worthy of the efforts of able scientists, and the problem does receive effort and attention. There is a vast amount of deuterium in the water of the earth, and we know that if we could burn this deuterium in a nuclear fusion reactor we would have virtually limitless energy. We would in a sense be imitating our own sun.

Much effort will be required to bring into operation the first successful fusion reactor. When it does succeed, and there is a high probability that it will, as the rate of progress during the past 20 to 30 years has been very great, we should have an energy source that presents fewer environmental problems than a typical fission reactor. The hazards will not, however, by any means be entirely negligible. The principal one will arise from the very large flux of neutrons that will be produced during the nuclear burning. Neutrons penetrate matter rather readily, but they can be absorbed, and it can be arranged that this absorption does not give rise to very great amounts of radioactive material. Thus, the problem of radioactive waste, which presents major difficulties in fission reactor plants, should be largely eliminated. There will be some difficulty in dismantling fusion reactors for repair, even if the problems are much less than those experienced with fission reactors.

Perhaps enough has been said to establish that energy sources available to man look promising, and that we can be optimistic about the future. There may be no guarantee as yet of an almost limitless amount of available energy, as this really awaits the successful outcome of fusion experiments, but for several hundred years the known available resources should suffice for human needs. This does not mean the needs will definitely be met, because there are always vexed questions of cost, availability of supply, and distribution.

## Energy conservation

The present pattern of consumption of energy in the advanced countries is discussed in detail in chapter two. Results indicate that in a country such as the United States, roughly one third is used in manufacturing industries, one third is used in supplying domestic needs, and about one third in transport. The amount per capita is very high indeed. If the same amount was used by other countries, we would need to provide about seven times as much total energy as is required at present. This is a daunting requirement, and could well be beyond the capability of scientists and engineers.

It is vitally important to find an answer to this problem. One answer that seems feasible, or at least could go a good way towards providing the total answer, is the possibility of greatly increasing the efficiency of utilization of energy. We noted that most power plants and also most machines have an efficiency of less than 40%; diesel engines are usually operating at under 20%. If we could win a factor of two by increasing efficiency, this would represent a very notable advance.

This is not beyond realization. We must realize too that countries like the United States are much involved in industries that require unusually large amounts of energy, for instance, steel making. Not all countries need to be involved in such manufacturing processes. Nevertheless, it should be added that it will prove incumbent on the people in countries which are large consumers to do everything possible to reduce their needs, as well as to devote all available effort towards more and more efficient use of the energy consumed.

Are there other ways of restricting consumption? One obvious way is to pay more attention to the problem of waste of material which costs energy to produce. Thus efforts to bring iron and steel, aluminium and copper, back into manufacturing as scrap could save a great deal of energy. Other metals and materials could likewise be conserved with profit.

While there may be methods of reducing consumption, we have to ask if there are any future needs which may involve us in increasing consumption. It seems unlikely that there will be, unless man is going to squander energy. In transport, effort is already directed towards greater economy of fuel, with the possible exception of trans-sonic flight, but even here it is clear that there is little difference in the energy consumption per passenger-mile. Those who argue against aircraft such as Concorde do so on environmental considerations. It has been maintained that the exhaust gases of such aircraft at the very high altitude at which they fly will destroy some of the ozone in the upper atmosphere, and that this removal could have serious weather effects. Others maintain that this is not so. The rate of introduction of such aircraft is likely to be low, so the question may be answered in practice before it becomes a matter of concern or indifference.

It is impossible to foresee what scientific and technological goals man may envisage in the future. We can guess that most of his new projects will require large amounts of energy. Perhaps the conquest of space is typical of the type of venture that is in mind here. The production of liquid hydrogen to power such vehicles requires large amounts of energy. It would only be necessary to worry about such a project if space travel were much more commonplace than seems likely at present.

### The population problem

Among the more disturbing prospects ahead is the possibility of uncontrolled growth of the human population on earth. There would be a temptation and an actual necessity to boost very considerably the production of all kinds of agricultural and horticultural products, and livestock would be required for meat production in greatly increased numbers. We now realize that the production of food can be costly in energy, and there could be very grave difficulties ahead if there were a population explosion. What will happen is a matter of opinion; optimists incline to the view that man is a thinking animal, and that he will not expose himself to such a dangerous situation.

It must be agreed that there are many experts who take a gloomy view of the population question. They consider that the population explosion in the less advanced countries will inevitably continue into the future, and raise the most daunting difficulties man has yet encountered. The methods of birth control have had less success than was thought necessary and desirable, and indeed in many countries the effect so far has been marginal. The linkage that exists between birth control practice and such matters as education, standard of living and religion, to name a few aspects of life bearing on the problem, is obscure. Thus the final solution is not clear. The optimistic expert may point out that in the advanced countries population can remain stable. It has remained stable in certain of the developing countries for hundreds and even for thousands of years. The impact of modern medicine has, however, brought about a sudden change in the average life expectancy, with the result that within a generation or so one of the main factors controlling population growth has altered completely. The life expectancy has advanced within several decades from approximately 25 years, in at least one country, to well over twice that value. In effect we could foresee a doubling of the population in such a country within less than fifty years, and this in itself presents a major problem. If the number of offspring of the average family is considerably above two, population growth occurs at a rate which could prove unmanageable; a country where this situation obtains would certainly be in grave difficulties.

The effect of population explosion in creating a major transformation as regards energy needs is clearly a subject requiring urgent careful study. We have seen that the average consumption of energy in advanced countries imposes severe supply problems at present. If a population, say ten times as large as that of the major consuming countries, had to be supplied on anything approaching the scale presently adopted in the advanced countries, the system would perhaps prove completely incapable of responding adequately. In this event we have to take into account the possibility that the standards of living in the developed and developing countries become divergent and not convergent, and that the disparity becomes even more marked than at the present day.

Energy and population are therefore closely linked, and control of population could become as critically important to the future of man as energy supply. The two matters are so closely linked that wise statesmanship and moral considerations will necessarily be required in reaching the right solution. In previous ages subjects of much less far-reaching consequence have resulted in major wars. The challenge to man and his knowledge is very clear, and it is imperative that the correct solution to the problems be found within a very few decades. Energy supply has become a matter of world-wide interest, and inevitably it will continue in the years ahead to be a subject of major importance to every nation on earth. earth.

This interplay of energy and population which has just been explored brings us to stress the vital importance of finding a truly basic means of providing energy abundantly and on an extremely long time-scale. We have shown that there is no guarantee that man can achieve this goal. Apart from finding what might be termed a partial answer, in the direct use of solar energy and the conversion of such energy, say by truly successful development of the solar cell, the best possibility lies in the release of fusion power. No one can be at all certain that fusion can be realized on a practical scale in a controlled thermonuclear reactor, so research effort on this has been increased in recent years. There is still much to be done in this field, and meanwhile no satisfactory long-term answer to the energy problem is available to us. However, the fossil fuels and other sources seem capable of supplying almost all our needs, in spite of the population explosion, for a few hundred years. In that sense there is no crisis, but the fuels have to be provided and the energy released and distributed at a price that can be afforded, or there will doubtless be regions of the earth in deep crisis at various times.

**A look at the future**

We have speculated in respect of the long-term abundant supply of energy, but we have not discussed to any extent the question of entirely new requirements for energy that could possibly arise. We saw that the consumption of energy in transport had grown greatly during this century, and it is appropriate to ask if other needs could likewise cause a steep increase in energy demand. Certainly in some advanced countries the control of the enclosed environment has brought about a steeply increasing energy requirement. Thus, in cities like New York, air-conditioning forms a large part of the consumption, both in the domestic scene and in the industrial/commercial scene. If this standard of control of the environment became world-wide, it would mean that a greatly increased consumption would obtain. To help us to answer the question more specifically, we remind ourselves that our surroundings, whether in home or factory or office, are important; our food is important, and here again we saw that agriculture could be very dependent on energy for best results, and our clothing is one of the important fundamental needs. In clothing, many have gone from fabrics based on natural materials to man-made synthetic substances; while this can involve additional energy need, it is not likely to prove a major stumbling block. To speculate beyond everyday requirements is to enter the realm of prophecy, and perhaps this could be an unprofitable exercise.

Industry continues to demand more and more energy, but we have looked at the matter of the steady rise in consumption. To the question of the possibility of a step-like increase, no simple answer is readily forth-coming. All that can be said is that there is no obvious reason to expect a sudden discontinuous increase in demand for energy.

There is one aspect of the energy scene of the future that deserves closer scrutiny. We have not considered the alternative to a population explosion. Yet population decline would be the logical step to encourage. Standard of living and energy consumption on a per capita basis are more or less linearly related. If energy consumption increases in industry, this is largely because more and more sophisticated machines are introduced, and these can be regarded as replacing operators of more simple machines. We are not saying that machines displace men but, as industries become more and more efficient, there can be a difficulty in maintaining full employment. If the population is allowed to decline by natural wastage and improved birth control practice, the standards of living can be raised and the total energy demand reduced or held down. Without doubt many other intrac-table problems arising from a continuously increasing population and an improving life expectancy are more readily resolved. Perhaps it is fair to emphasize that environmental problems including that of pollution of the atmosphere and biosphere are also reduced in severity. In the energy study we have certainly good scientific grounds for advocating the con-tainment of populations all over the earth. At the same time we realize that here we encounter severe religious, moral and political questions. In some parts of the world men have survived for thousands of years in conditions of extreme energy want, but usually without population increase beyond a balanced stage. There are few fundamental limitations giving this population balance today; in this respect there is a new situation, a situation in which man himself may have to make a deliberate choice. This choice is one which can be expressed simply as between large family and poor standard of living, and small family and high standard of living. There is little clear evidence that such a choice has yet been made, but optimistic examination of the data might seem to support the view that the smaller family and the higher standard of living is preferred. It is often forgotten that the large family was intended to safeguard against want and poverty, but increasingly it tends nowadays to bring poverty and want.

There were many conditions of life in the past which consciously and subconsciously led man to favour the larger family. Life was hazardous, and mutual support within the small community was at times absolutely essential for survival. Life expectancy was very low, and the larger family was in some senses a form of insurance. In recent times and even in the less advanced countries, the constraints and hazards to life have been eliminated or greatly reduced. In such circumstances family planning ideas can be propagated successfully. On such lines the scale of the energy and environmental difficulties in the years ahead can be seen to be lessened very considerably.

## Avoiding waste

In a thoughtful society much attention would be paid to energy conservation. By this we mean the expenditure of effort in the more efficient use of energy. Inefficient modes of use are easily found. In large power-generating plants, about 60% of the input energy is dissipated in cooling gas, or vapour, or liquid. These cooling materials are very seldom used for low-level space heating, but are simply wasted. Indeed in most homes, offices and workshops, the space heating is seldom tackled in a scientific fashion with maximum accent on energy saving. Walls are generally poorly insulated, and the heat loss on that account can be very substantial indeed. If much more thought and effort were devoted to the proper efficient utilization of energy in its many applications, large savings could be effected. Without doubt there is unjustifiable waste of energy on a large scale, and it is to be hoped the future will see vast improvements introduced. This general problem is discussed in chapter three.

We recall that one of the fundamental laws of physics is that energy cannot be destroyed. It can be converted from one form to another. The potential energy of a body at a height above ground can be converted to the kinetic energy of the same body as it falls to earth; the thermal energy of a fuel can be converted to electrical power, heat in the cooling liquids, etc. In talking of waste of energy we are not at all implying that energy is literally destroyed or lost. What is meant by the phrase is that the energy is in a degraded form; the materials in which the energy is present are, say, at a temperature which is so relatively low that we cannot make effective use of the energy. We are witnessing the second law of thermodynamics at work. Order tends in nature to increasing disorder; when we 'waste' energy, we mean it is in a state from which we can only with difficulty recover some of it for useful purposes.

Much of the energy in our universe has its origin in the burning stars of which our sun is a fairly typical example. It pours out into space an immense amount of thermal and other types of radiation. In equivalent electrical energy it has the fantastic output of around $10^{20}$ megawatts, and the small fraction reaching the earth provides in favourable atmospheric conditions about 1 kW per square metre. Most of the radiation of the sun moves out into 'cold' space. The average temperature of outer space is only several degrees absolute, so cold in fact that no form of life could survive. Our planet, earth, happens to be in the near vicinity of our star, the sun, and we are bathed in the radiation falling on the surface of our earth. It is for this reason we can live and do much more than survive. If we can reproduce on earth the fusion process of our sun and burn some of our own hydrogen, we shall in imitating the sun (on a comparatively minute scale) be in a position to provide ourselves with huge energy resources for many millenia.

## Environmental and sociological problems

Perhaps it is appropriate to mention once again the constantly recurring problem of the abundant production of energy required to meet the wide range of human needs, and the adverse effects on environment that can stem from a comprehensive programme. Without doubt the full answer to this dilemma has not been forthcoming, and it might prove impossible to provide a solution to satisfy all aspirations. The answers provided by scientists, economists, sociologists and politicians will be at variance with each other. Scientists are generally in favour of energy sources of long life, but are no better qualified than others to decide on the balance between abundance of supply and, for example, health and safety of operators and the public at large. Economists will be mostly concerned with advising on the economic aspects of operating one source compared with others, and in discussing and comparing sources insofar as their impacts on the wealth and prosperity of society as a whole are concerned. Conversely, politicians and sociologists are likely to be concerned with employment prospects, the generation of prosperity, and the health and well-being of the maximum number of people. They may be more inclined to provide well in the short-term and disinclined to put much stress on long-term prospects. This is not to say that they are in any sense wrong in emphasizing the shorter-term needs. Some scientists and economists hold (and offer evidence to support) the view that truly long-term answers are elusive, and demand an ability to prophesy that is not likely to be forthcoming. This inability to see far ahead is observed in many spheres, and scientists appreciate that scientific discoveries have a tendency to emerge in almost unpredictable ways.

If we were to accept this kind of argument, we would not be too anxious about some of the problems that seem rather alarming at the present time. We have, for example, no perfect answer to the difficulty of storing the large amounts of radioactive waste in the United States, where at present between fifty and sixty power reactor stations are in operation. Huge tanks are used to store this waste, while scientists and engineers seek a more thorough solution. There is a search for a large undisturbed salt deposit in which the waste could be stored at some depth in the earth, but a suitable one has yet to be found. This has caused environmentalists much concern, which will not be reduced till a better answer is forthcoming.

It should not be assumed that fusion reactors will provide an answer to the environmental problems posed by the introduction of large numbers of power-generating fission reactors. The huge flux of neutrons liberated in very-high-temperature machines such as fusion reactors will have a considerable effect on containment vessels, and scientists have still to explore the nature of some of these effects. We know comparatively little about the physical and chemical changes produced in ceramics, metals, alloys and construction materials of all kinds. However, it must be stressed that it is impossible to operate fission reactors without producing large amounts of dangerous radioactive elements of all kinds, some of long life and hence potentially more hazardous. In addition to these radioactive substances which are produced in the act of fission itself, plutonium is produced in large amounts, and this element, of half-life 24,000 years and alpha-emitting, presents particularly difficult problems. Ingestion and inhalation of very minute amounts could prove dangerous, indeed fatal. No equally serious problems seem to arise in the case of fusion reactors, although tritium, which is a heavy radioactive form of hydrogen, will be present. Indeed the main reaction in the machine is likely to be that between deuterium and tritium, giving rise to helium and a neutron; released neutrons acting on lithium will, in turn, provide tritium for burning. To the scientist these main reactions are essentially 'clean' compared with fission, and the final products can be almost wholly stable. Tritium has a half-life of 12 years, which is comparatively short, and does not present nearly so serious a threat as plutonium.

We conclude by pointing to fusion as a possible scientific answer to man's energy requirements, and one that has great potential duration of supply. It appears to offer many attractions, when viewed from the standpoint of those concerned with the environment, human health and well-being. The scientist has yet to show that the fusion machine is practicable, and so considerable ingenuity and skill is still required. This situation of challenge is one familiar to man. Those of us who are optimists believe that success will come and that the energy needs of man will be supplied in a satisfactory way.

## FURTHER READING

Bragg, W. (1929), *Concerning the Nature of Things*, Bell. Intended for the intelligent reader, this short volume deals with the fundamentals of physics.
Gamow, G. (1952), *The Creation of the Universe*, Mentor Book of The Viking Press Inc. and Macmillan & Co. Ltd. A remarkably lucid account of this fascinating topic.

Hoyle, F. (1950), *The Nature of the Universe*, Blackwell, Oxford.

Bondi, H., Bonnor, W. B., Lyttleton, R. A., Whitrow, G. J. (1960), *Rival Theories of Cosmology*, Oxford University Press.

Gass, I. G., Smith, P. J. and Wilson, R. C. L. (editors) (1972), *Understanding the Earth*, Open University Set Book, Science Foundation Course, Artemis Press, 2nd edition. A fascinating, if rather difficult, book with chapters covering geothermal energy and other pertinent topics.

Woolner, A. H. (1968), *Modern Industry in Britain*, one of 'The Changing World' series of the Oxford University Press. Excellent as an indication of the extent of industry's dependence on energy.

Gregory, D. P. (1972), *Fuel Cells*, Mills and Boon Ltd., London. A more technical treatment of one of the problem areas in energy studies—the effective storage of energy.

*The Petroleum Handbook* (1959), Fourth Edition, of Shell International Petroleum Co. Ltd. Technically-based but extremely readable in many important sections.

Allis, W. P. (editor) (1960), *Nuclear Fusion*, D. Van Nostrand Co. Inc. Gives, essentially for the more scientific reader, an excellent story of the earlier work on nuclear fusion.

Perlman, R. and Connelly, P. (1974), *The Politics of Scarcity*, Oxford University Press. Provides some of the thought-provoking questions regarding natural resources in general.

Smyth, H. D. (1945), *Atomic Energy for Military Purposes*, Princeton University Press. The official US account of work on nuclear power reactors and bombs is historically important. Many volumes have appeared since.

Groueff, S. (1967), *Manhattan Project*, Collins.

Kramish, A. (1960), *Atomic Energy in The Soviet Union*, a Rand Corporation publication, Stanford University Press. Traces nuclear and development work in the USSR.

# CHAPTER TWO

# FLOW OF ENERGY THROUGH TECHNOLOGICAL SOCIETY

EARL COOK

## How much?

An industrial nation uses a lot of energy. For comparison with nations or societies which are not industrialized, the rate of per capita energy consumption is commonly used (Table 2.1). When only fossil, nuclear and falling-water energy sources are considered, dominantly agricultural countries fall generally below the world average of 38,200 kilocalories* per person per day (1971), whereas industrial societies range upward to more than 220,000 (figure 2.1). The per capita consumption of such energy in the developed countries is more than 17 times that in the undeveloped countries.

Among the countries considered industrialized, we see a wide range in per capita energy consumption; the United States, for example, consumes for each inhabitant more than three times as much energy as does Switzerland or Japan and twice as much as the United Kingdom. Does this mean that the average US citizen is that much more happy, healthy, and comfortable? Not at all. Wide differences in per capita energy consumption among industrialized nations reflect several factors, among which are efficiency of energy use (US automobiles are less than half as efficient as those in the rest of the world), mean transport distances (the United States is a big land), the magnitude within the national economy of energy-intensive activities such as mining and manufacturing relative to business and other service activities, the extent to which comfort heating and

* 1 kilocalorie = 4·18 kilojoules.

30

**Table 2.1**    1971 Per Capita Energy Consumption*

| Country | Daily per capita energy consumption (kcal) | Percent (net) imported |
|---|---|---|
| USA | 222,400 | 10·6 |
| Kuwait | 201,900 | NE |
| Canada | 184,600 | NE |
| Czechoslovakia | 131,000 | 18·6 |
| Bahrein | 127,600 | NE |
| East Germany | 124,900 | 21·5 |
| Belgium | 121,000 | 94·6 |
| Sweden | 120,500 | 89·5 |
| Developed Countries (Average).........118,900 | | |
| | | |
| United Kingdom | 109,000 | 46·8 |
| Australia | 107,800 | NE |
| Denmark | 105,500 | 100·0 |
| West Germany | 103,400 | 50·4 |
| Norway | 103,200 | 60·9 |
| Netherlands | 101,400 | 36·9 |
| USSR | 89,800 | NE |
| Poland | 86,500 | NE |
| France | 77,700 | 78·1 |
| Switzerland | 70,800 | 80·0 |
| Japan | 64,600 | 98·1 |
| South Africa | 57,300 | 27·0 |
| New Zealand | 55,800 | 58·0 |
| Israel | 52,600 | 0·1 |
| Italy | 52,600 | 91·1 |
| | | |
| World Average.........38,200 | | |
| | | |
| Argentina | 34,900 | 11·7 |
| Spain | 33,000 | 82·9 |
| Greece | 29,100 | 69·8 |
| Mexico | 25,100 | 3·3 |
| Iran | 20,300 | NE |
| Portugal | 16,100 | 100·0 |
| Peru | 12,200 | 29·6 |
| China | 11,100 | 1·5 |
| Brazil | 10,200 | 53·7 |
| Undeveloped Countries (Average).........6,900 | | |
| | | |
| Egypt | 5,600 | NE |
| India | 3,700 | 18·1 |
| Indonesia | 2,500 | NE |
| Pakistan | 1,600 | 51·5 |

NE = net exporter

* Fossil and nuclear energy and hydroelectricity; food, fuelwood, animal and human power not included.

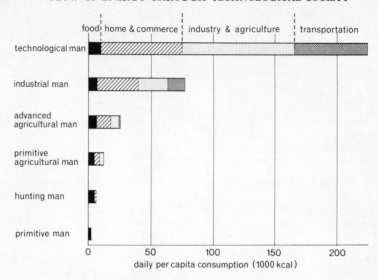

**Figure 2.1**  Man's daily energy consumption at several levels of socio-economic development.

cooling is practised in residential and commercial space, and the amount of energy spent in military activities and material.

Quantitative aspects of the subject of this chapter will refer mainly to the United States, because that is the nation whose energy economy the author knows best. It seems to be assumed frequently that the US example will inevitably be repeated by other industrial nations as they continue to 'develop' economically. This is not at all inevitable and, indeed, present signs indicate that the energy consumption of the United States will decline over the next few decades, and that the patterns of energy use in those industrialized countries which are able to retain access to adequate energy resources will become more alike, at a gross consumption level more like that of the United Kingdom today than that of the United States; there will be very strong emphasis on efficiency, on getting more useful heat and work out of more costly energy-delivery systems. As we shall see, the US energy system is a wasteful system, and for that reason alone can be an instructive example of what industrial society may no longer be able to afford.

Energy enters the US system mainly in fossil-fuel form (figure 2.2). The several sources contributing in 1971 are shown in Table 2.2.

Electricity is such a clean and adaptable form of energy that its use has been growing rapidly. In the United States, more than a quarter of the gross energy consumed is used to generate and transmit electricity. In that process about 20% of the total energy input to the economy is lost as inutile heat.

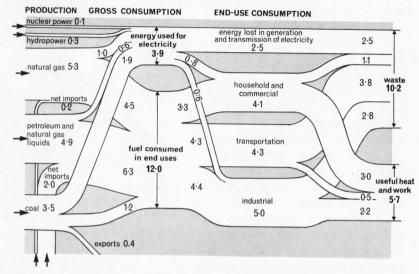

Figure 2.2    Approximate flow of energy ($10^{15}$ kcal) through the US economy, 1971 (Cook, 1973).

Table 2.2    Sources of Energy Contributing to the US System in 1971

| Source | Amount ($10^{15}$ kcal) | Share (%) |
|---|---|---|
| Petroleum and its products | 6·9 | 43 |
| Natural gas and natural-gas liquids | 5·5 | 35 |
| Coal | 3·1 | 20 |
| Hydroelectricity | 0·3 | 1·5 |
| Nuclear electricity | 0·1 | 0·5 |
|  | 15·9 | 100 |

The electricity generated and that portion of the fuel input not used in electricity generation are used in the four major sectors of the economy approximately as in Table 2.3.

Table 2.3    Use of Energy in the US Economy in 1972

| Sector | Amount ($10^{15}$ kcal) | Share (%) | Annual growth rate, 1960–68 (%) |
|---|---|---|---|
| Industrial | 5·0 | 37 | 3·9 |
| Transport | 4·3 | 32 | 4·1 |
| Residential | 2·3 | 17 | 4·8 |
| Commercial | 1·8 | 14 | 5·4 |
| Totals | 13·4 | 100 | 4·3 |

(Stanford Research Institute, 1972)

In the end-use of energy, further generation of inutile heat takes place, resulting in an overall efficiency for the US energy system of no more than 36%.

These matters will now be discussed in greater depth and detail.

## Where it comes from

### The fossil fuels

**Coal**    Modern industrial society has been built by burning the fossil fuels, and it is at present utterly dependent upon the continued availability of the fossil fuels at or above present rates of consumption. The fossil fuels most used are crude oil, coal, and natural gas. These are non-renewable resources, representing solar energy of the past converted by photo-synthesis into plant material and subsequently, in both vegetable and animal form, gathered into geologic traps by physical processes and turned into hydrocarbon concentrations having high stored-energy values. Not only are the fossil fuels not renewable—at least on any time scale signi-ficant to man—but they cannot be recycled or re-used, because they are destroyed during combustion. On the basis of this ephemeral source of inanimate energy, first the peoples of western Europe, and then those of other regions possessing or having access to fossil-fuel supplies, have greatly increased their numbers, their life span, their material standard of living, and their perturbation of the physical and biological environ-ment.

Marco Polo reported that the Chinese burned black rocks, and they are known to have made use of natural gas almost a thousand years ago; but the Chinese never made these fossil fuels the base of an industrial society; only recently has that great sluggish dragon commenced to stoke its fires with coal and shale oil.

Independent discovery of the phenomenon of the burning rock was made sometime early in the present millennium on the north-east coast of England. Here, as in China, the discovery might have led only to local and fitful use of coal had not the Elizabethan energy crisis intervened. Forests, expecially in the southern half of England, were over-harvested for ship timbers, for firewood, and for charcoal manufacture for England's growing metallurgical industry, so much so that London (by the time that Elizabeth had outwitted and outlived her elder sister Mary) was hard up for fuel for its many hearths. Coal was brought to London in schooners from the Tyne, but the stinking pall it laid over the city caused it to be banned for most of the period during which the royal family and much of the powerful gentry still lived there. In Elizabeth's day, however, the need became so pressing that a compromise with environmental quality was effected; the ruling class established new residences outside the urban smog zone and lifted the ban on coal, whereupon the flow of dusty schooners from the north-east coast became a flood; the remaining forests could be devoted to shipbuilding, charcoal-making, and hunting preserves, while the common citizens of London, as John Evelyn so mordantly limned them in *Fumifugium* (1661), choked, hacked, spat, and died of phthisic disorders.

Coal was mainly used for comfort heating and relatively low-temperature process heating until Abraham Darby perfected in 1709 a method of making from coal a coke that would perform better than charcoal and thereby opened the Age of Steel. Thenceforward, iron and coal were the twin djinns of the Industrial Revolution; the use of both expanded enormously and still today, although new types of furnaces are being used, the production of most of the world's steel depends upon a supply of coke made from 'metallurgical-grade' coal.

No geologic resource is uniformly or even randomly distributed throughout the world. Some areas and nations have abundant resources of one or more such resources while others have none. Coal is no exception; it is strikingly concentrated in the temperate zone of the northern hemisphere, and it is no accident of coincidence that the principal industrial nations of the world are those favoured by nature with large coal deposits—as well as iron ores.

Coal not only allowed a faster and cheaper means of producing steel, but also a source of power far greater and more reliable than that previously available from wind, falling water, and muscles. It offered an attractive alternative to firewood in those regions where firewood, human labour, or both, were scarce. Only in those areas where fuelwood was abundant and human labour cheap did the use of wood for firing locomotives, steamships and stationary boilers persist into the coal era.

Power is needed for speed. For more than 4000 years the limits of speed in communication and transport had been the speed of a running horse and the speed of the swiftest sailing vessel. Coal and the steam engine changed that. The steam engine was developed in England to pump water from the deepening coal mines, and was adapted in the locomotive to haul coal from the mines to the docks. This part of history laid the foundation for the great edifice of modern industrial society. Soon the coal-fueled steam engine would power trains, ships, and mills of many kinds; would hoist ore from the depths of the earth, drag logs from the virgin forest, and drive pilings to protect ports and bridge canyons; finally, it would turn generators to create electricity, that clean genie of the technological world.

In the United Kingdom and Europe, coal comes mainly from underground mines, but in the United States more than half of the coal mined is taken from surface mines. Underground mining of coal is costly and hazardous compared to surface mining. On the other hand, it appears to cause less environmental damage; but this depends on the extent to which surface-mined land is rehabilitated for other uses, and upon the extent to which underground mining produces long-delayed effects such as surface subsidence and acid mine drainage.

**Crude oil**     Coal, with all its enormous benefits, has some drawbacks. In addition to being dirty and creating air pollution when burned, coal is not as versatile as one might wish. For example, a coal-burning automobile, although quite feasible technically, would be cumbersome and an all-round nuisance. A coal-burning aircraft might be even more troublesome. With coal there is always ash and clinker to be disposed of. The need to use steam as a means of converting the chemical energy of coal into mechanical energy introduces further constraints of efficiency, maintenance costs, and accident risks.

Crude oil and its products can surmount some or all of these problems of coal. Fuel oil is cleaner, easier to handle and to store than coal. The internal-combustion engine, unlike the steam engine, is effective and instantly available in a wide range of sizes. Indeed, it was the development of the internal-combustion engine, mainly in Germany in the closing years of the nineteenth century, its application to the automobile, mainly in France and the United States during the next few years, and the availability of abundant, cheap, clean, and powerful fuel—gasoline—which led to an upsurge in the use of crude oil and to the present dominance of oil in the energy economies of all industrial nations.

For thousands of years it has been known that crude oil, either fluid from natural seeps, or congealed as asphaltic lakes, would burn. Asphalt was used as a fuel in Mesopotamia as early as 3000 B.C. But oil had to wait even longer than coal for the development of effective exploitative technology. The first large-scale use of a crude-oil derivative was for lighting. Kerosene lamps replaced whale-oil lamps in the middle of the nineteenth century as the price of whale oil rose because of increasing demand and limited supply. The first successful oil well known was completed in Romania in 1857. During the first century of modern oil exploitation, supply has persistently exceeded demand; consequently, the price of oil and its products has been low, except during those times when producers' cartels have been effective in maintaining a non-market price level.

Demand for kerosene was small until the invention of the Welsbach mantle in 1886 tripled the amount of light obtainable from a drop of that aromatic fuel; thereupon the demand also tripled. But still demand was feeble in the face of abundant supplies. The proliferation of the automobile in the early years of the present century greatly increased demand, but new oil discoveries such as Spindletop (Texas) in 1901 and the great East Texas Field in 1932 kept supply well in the lead and prices low. Consequently, automobiles and trucks could be designed and marketed with only modest thought given to fuel economy and none whatsoever to fuel

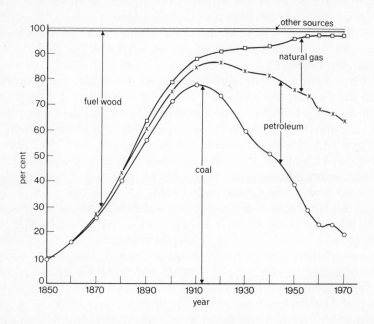

**Figure 2.3**   Input of energy into US economy, 1850–1950. Not shown in the figure are hydropower (1·1% in 1950, 1·4% in 1970) and smaller sources—human workers, waterwheels and work animals (0·94% in 1850, 0·08% in 1950).

availability. In 1900 there were 8000 automobiles in the United States; in 1920—8,000,000, and in 1971—94,000,000. But oil discoveries kept pace with burgeoning demand.

In the early years of the present century, fuel oil displaced coal as a fuel for steamships; shortly after mid-century, a similar replacement took place in locomotives. Since about 1920, the growth of the world fleet of aircraft has contributed to crude-oil demand, but only since about 1940, with the great stimulus of World War II, have aircraft become significant consumers of oil products. In the United States, the replacement of coal by oil took place largely in the 1920–60 period (figure 2.3).

Today, the industrialized world is heavily dependent on oil, and the physical and political control of oil is the dominating concern of international relations. Even Sweden, richly endowed with hydroelectric power potential, now imports more than half its total energy needs in the form of crude oil and its products (Table 2.1). Japan, without abundant domestic energy resources, imports 98% of its energy needs—mainly as oil, but some as metallurgical-grade coal.

Still the world's oil-producing capacity continues to grow faster than demand. Especially is this true now that the Organisation of Petroleum-Exporting Countries (OPEC) is able to maintain the world price far above the cost of production and the price of a few years ago. The present price is restricting demand.

Oil is somewhat more broadly distributed throughout the terrestrial world than is coal. However, 85% of known world reserves (figure 2.4) are in the hands of OPEC, representing less than 8% of the world's land area (and about 6% of the global population). Only four countries have both coal and oil in large amounts: the Soviet Union, the United States, Canada, and China. Of these, both the United States and Canada are far along toward exhaustion of oil reserves, but retain abundant coal reserves; the Soviet Union and China, on the other hand, are still finding and developing large new oil deposits, while the remaining reserves of coal, which they have made relatively more use of than oil, are substantial.

Oil is more easily and cheaply transported than coal. It can be made into a great variety of products, only some of which are fuels. Its extraction from the earth—despite the highly publicized Santa Barbara oil leak—involves less environmental damage than does extraction of coal. In its transport by enormous tankers, however, there is much more risk to the ocean and littoral environments than in any transport of coal. The greatest environmental impact of both fuels follows combustion, and consists of the spewing into the air, from multiple point sources, of carbon monoxide, sulphur and nitrogen oxides, unburned hydrocarbons and ash.

**Natural gas**    Natural gas often is discussed with crude oil, as if the two were inseparable companions, at least in nature's vaults. They are not. Although perhaps formed together, they may not lie together. Both are fugitive materials, the gas the more so than the liquid. For this reason, it may escape traps oil cannot and may exist in reservoirs which oil will not penetrate. In Siberia there are very extensive natural-gas fields with little or no associated crude oil. In general, however, world-wide distribution of natural-gas reserves is similar to that of crude oil. Most are in countries also rich in oil (figure 2.4).

For many years, natural gas produced with oil was wasted, because there was no way to carry it to the urban markets. World War II stimulated the construction of long-distance pipelines, not only for oil, but for natural gas. In recent years, development of liquefaction technology and cryogenic tankers has changed the 1940 picture dramatically; now no natural gas

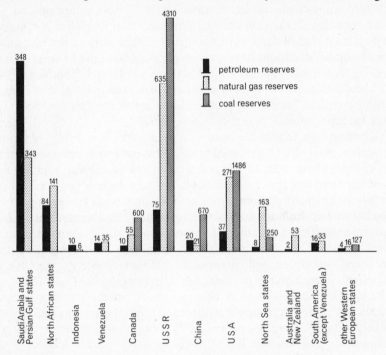

**Figure 2.4** Fossil fuel reserves (barrels × 10⁹) of selected nations and groups of nations. Petroleum and natural gas reserve estimates from *Oil and Gas Journal* (1972). Coal reserve estimates based on Averitt and Paul (1969).

need be regarded as a waste product. Because most of the impurities in natural gas are removed easily at or near the wellhead, the delivered product burns cleanly. This inherent cleanliness and the ease of handling, once pipeline systems are installed, have made natural gas a premium fuel.

*The nuclear fuels*

In 1940 the first controlled fission reaction took place. Since then controlled fission of the only naturally fissionable radioactive isotope, uranium-235, has been developed for both military and peaceful uses. The peaceful uses of atomic energy are: (1) for conversion into electricity, (2) for excavation of harbours and canals, (3) for shattering of rock masses to liberate natural gas or to allow the introduction of leaching or heat-transfer fluids incident to metal or heat recovery, and (4) fusion of a deep mass of rock to encapsulate highly noxious waste materials. Only the first has proceeded beyond feasibility studies and prototype testing; only it will be discussed here.

Uranium is not found in the same kinds of rocks as are oil and gas. Consequently, it is not surprising to discover that its global distribution differs from that of the fossil fuels, being similar only in the fact that a small number of nations have a very large portion of the known reserves; among these nations are the United States, Canada, and the Soviet Union.

The development of nuclear electricity has been much slower than its enthusiasts had hoped and predicted, but much too fast for those who fear its environmental and social consequences.

The naturally fissionable isotope, uranium-235, is but 0·7% of natural uranium, most of which is uranium-238. Present power reactors are able to convert about 1·5% of the potential energy of natural uranium into kinetic energy and then heat; in the reactor core some of the uranium-238 is fissioned by neutron bombardment, and contributes energy in addition to that derived from the fissioning of uranium-235. Thorium can also be used in a reactor (with uranium) to produce more energy than the uranium itself will yield. So-called breeder reactors, now in the development stage in several countries, will increase the amount of energy that can be obtained from natural uranium (or thorium) by 60-fold. They will eliminate concern about there being adequate fuel for a great expansion of nuclear power, replacing it with concern for the safe management of highly radioactive wastes, as well as of the extremely toxic plutonium, a manufactured fissionable fuel that will be much used in breeder power plants (Cook, 1975).

The breeder power-plant appears certain to proliferate as industralized nations seek alternatives to the fossil fuels. Nuclear electricity, which today contributes a minor amount to the energy economy of the industrialized world, may account for 25% of the energy budget of a high-energy nation early in the coming century (Cook, 1972).

There can be no monopoly on nuclear technology. Bombs and reactors can be built by anyone with the materials and the capital. Bombs are cheap; nuclear power-plants are expensive. A few bombs take relatively little radioactive material; large power reactors supplying a high-energy society over decades will take relatively more. The constraints on making (and using) nuclear explosives thus will be less than on building (and using) nuclear power-plants. Nuclear electricity is now considerably more expensive than electricity from large modern conventional thermal power-plants (fuelled by coal, oil, or natural gas). The advent of the breeder-reactor power-plant is not likely to reduce the cost of nuclear electricity. It will increase uranium reserves enormously and allow stockpiling of many years' supply of fuel, so that interdiction of supply can be greatly reduced as a threat to nations not possessing domestic uranium or thorium resources. A large nuclear-power industry will require a large amount of capital investment and assured availability of fuel.

The environmental hazards can be kept low at the power-plant by proper engineering and siting (preferably underground). They can also be kept low at the nuclear fuel-reprocessing plant (but have not been in the United States), again by adequate engineering and siting. Apart from the social peril involved in the potential for theft of weapons-grade nuclear material (either plutonium or uranium-233) in a breeder economy, the main environmental problem is waste management after the recovery of 'unburnt' fissionable material from the spent fuel elements.

Although expansion of the domestic nuclear-power industry has a high priority in the plans of most industrialized countries, there is as yet no generally-agreed 'safe' procedure for managing high-level long-lived (hundreds to hundreds of thousands of years) radioactive wastes. Disposal in ocean deeps seems hazardous in the long term, shooting the stuff into space is costly (and perhaps irresponsible), and storage in salt caverns or above-ground structures under perpetual surveillance appears quixotic.

No other energy system so plainly illustrates the theorem that the product of the economic process which accumulates indefinitely is waste. The work, warmth, and light furnished by nuclear electricity are ephemeral; the waste is not.

*Renewable energy resources*

**Hydroelectricity**    Placing electricity produced from the kinetic energy of water falling through a turbine under the heading of a *renewable* energy resource suggests the skewness of man's vision. Falling water impelling a waterwheel to do work *is* renewable. But even the largest waterwheel will not produce much electricity. What is needed is hydraulic head and discharge; the water must fall a considerable vertical distance, or there must be a great deal of it, or both. This need in most cases can be met only by creating a self-replenishing reservoir, usually by putting a dam in a river valley. Even natural lakes are transitory, being destroyed either by siltation or erosion of the outlet. Man-made lakes are more so. Because erosion of the outlet is not possible, and no large dam has yet been built with a satisfactory flushing arrangement, siltation proceeds apace. Many large dams will have a useful life of but 100–200 years, unless an economic means is found of removing many cubic miles of silt accumulated in the reservoirs. Consequently, hydroelectricity is not in truth a renewable resource. It appears, however, that only geologists, astronomers, and other historians see eternity as distant more than 100 years.

Like natural gas, hydroelectricity is a pleasant source or form of energy. It is clean, has relatively minor adverse environmental impacts, does not smell, and can be distributed to both small and large consumers through the same transport network. Again like natural gas, there cannot be enough to satisfy demand, for if the available unexploited potential were all developed, the contribution of hydroelectricity to the world's energy economy would still be less than 5% of the total; in the United States today, it is less than 2%.*

In both the USSR and the USA, hydroelectric power for several decades has enjoyed a mystique and has been propagated by a cult. Lenin and Franklin Roosevelt put the mystique into political form, so that it over-rode economics as well as social accounting; in both countries dams were built to produce hydroelectric power in the confident expectation of great secondary economic and social benefits, regardless of the locations of the dams and the nature of their geographic surroundings. In the United States, the bloom is off this rose, but it seems still fragrant in the realm where dogma dies a bit harder.

Possibly because of this mystique, the adverse environmental impacts of some large reservoirs were not foreseen and seem to have been perceived with shocked surprise (if at all) by the proponents of hydropower.

Induced earthquakes, provision of new breeding grounds for disease vectors, submergence of arable land, and the prevention of annual soil replenishment by flooding are among the major adverse impacts of some recent large dams and reservoirs.

**Other sources**    In countries such as Canada and the United States, which possessed, during the full flood of the Industrial Revolution, large forest resources, *fuelwood* was a large part of the energy economy for many years. As late as 1875, half of all US locomotives still burned wood. At present, however, wood is insignificant as a fuel, not because the forest has vanished (much of it has, of course) but because wood is an inefficient and expensive fuel in an industrial society; it requires too much human labour per unit of useful heat or work obtained. The same is true of *dung*, which once was the prized fuel of the American prairie (buffalo chips) and is much used today in non-industrial society; but in industrial society, feedlot

---

* Often given as 4%, because of the US Bureau of Mines' curious habit of calculating the energy equivalent of hydroelectricity as if it had been produced in a conventional thermal power-plant of 30%–33% thermal efficiency.

manure is regarded as an environmental problem, not as a valuable product.

*Wind* and *tidal energy* are of little use at present, mainly because of the difficulty and expense of storing the energy converted between the episodes of generation. *Solar energy*, apart from the ambient heat that makes possible all life as we know it, is much too expensive in converted form except for special uses. This is bound to change when fossil-fuel energy grows more expensive and as economies of scale and mass production are achieved. Solar energy for heating and cooling already is economic in some parts of the world and may soon become so in others. Solar energy for producing electricity, on the other hand, has a long economic-technologic row to hoe before it becomes viable for general use, if it ever does.

The *geothermal energy* now being used in Italy, Iceland, New Zealand, Japan, California, Siberia, and Mexico, is non-renewable, because it comes from pockets of heat trapped in the upper part of the earth's crust. If an economic way can be found to extract heat from dry rock at depths of 10,000–30,000 feet, geothermal energy will become almost a renewable resource.

## How it is used

*Proportions used for principal purposes*
In 1968 energy was consumed in the United States for the following purposes, in approximately the proportions indicated in Table 2.4; the annual rates of growth for the period 1960–68 are given also.

Table 2.4 is a simple statistical picture of an affluent industrial society,

Table 2.4    Energy consumption in the USA (1968)

| Purpose | Share (%) | Annual growth rate (%) |
|---|---|---|
| Process steam and direct heat in industry* | 27·6 | 3·3 |
| Transport | 26·2 | 4·1 |
| Space or comfort heating | 21·0 | 4·0 |
| Electric drive | 9·6 | 5·3 |
| Water heating | 4·3 | 4·5 |
| Air conditioning | 2·8 | 11 ± |
| Refrigeration | 2·8 | 5·6 |
| Lighting | 1·9 | NA |
| Cooking | 1·4 | 2·3 |
| Electrolysis in industry | 1·3 | 4·8 |
| Television and clothes drying in the home | 1·0 | 11 ± |
| | 99·9 | 4·3 |

* Includes cooking processes in industry. All figures re-calculated from information given in Stanford Research Institute, 1972.

in which inanimate energy at the point of use costs much less than human energy. Two further facts need to be known before Table 2.4 can be used to full advantage: (1) the US population growth rate during the period 1960–68 was less than a quarter of the energy-consumption growth rate, and (2) the cost of energy, especially electrical energy, in constant dollars ('real cost') decreased during that period. Consequently, it became possible to substitute energy derived from the fossil fuels not only for human work, but for clothing (air-conditioning) and play (television) as well.

### Industry

Although the industrial sector of the US economy consumes more energy than any of the other three sectors, energy consumption in industry has been growing more slowly than in any of the other three.

Industry produces goods. To produce goods, especially those whose basic ingredients must be won from the rocks of the earth's crust, requires more energy than to produce services; in other words to make steel products takes more energy than to buy and sell shares of stock. As an industrial nation matures, its internal market tends to become saturated, because the consumer is not an infinite sink. As many industrial nations mature, external markets likewise tend to become saturated, because in the non-industrialized nations there is a limited capacity to pay. Consequently, the economy of a mature industrial nation tends toward producing proportionately more services and less goods, because the capacity of the domestic consumer for services (health care, entertainment, sporting activities, education, drama, music, art, and various forms of pandering) remains greater than his or her capacity (or desire) for goods. This shift shows up in the partition of energy consumption; relatively more goes into the commercial and residential sectors than into industry. The share of transport may also increase if the motor-car employed as toy and phallic symbol cancels out the efficiencies of good transport by diesel energy. The share of industry decreases as industrial society evolves into technological society.

It happens also that the keen edge of competition keeps the use of energy in industry, in contradistinction to the use of energy in the *products* of industry, rather efficient and responsive to changes in energy costs. Consequently, industrial production may grow at a rate equivalent to that of the growth rate of energy consumption in the other economic sectors, while the energy consumed in that production, because of new economies achieved, grows at a lesser rate.

In the United States during 1947–66, both of these phenomena were at work, with the result that the gross national product grew faster than energy consumption (figure 2.5). In more recent years, that relation appears to have reversed.

Some industries use more energy than others. The differences are suggested by Table 2.5 which shows the main industrial user groups and their portions of the total US industrial energy consumption in 1968.

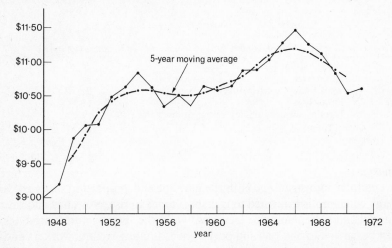

**Figure 2.5**   United States gross national product in 1958 dollars per million Btu gross energy consumption (Cook, 1972) (1 Btu = 1·055 kJ).

Table 2.5 suggests that at least half the energy used in US industry is devoted to extraction and processing of non-renewable geologic resources. The iron and steel industry accounted for 64% of the energy consumed by the primary metal industries; about 6% was used by the aluminium industry. Noteworthy is the absence of machine manufacture from the leading groups; if the energy used to manufacture all transport equipment

**Table 2.5**   Main industrial user groups and their portions of the total US industrial energy consumption in 1968

| Group | Share of total (%) |
|---|---|
| (1)  Primary metal industries | 21·2 |
| (2)  Chemical and allied products | 19·8 |
| (3)  Petroleum refining and related industries | 11·3 |
| (4)  Food and kindred products | 5·3 |
| (5)  Paper and allied products | 5·2 |
| (6)  Stone, clay, glass, and concrete products | 4·9 |
| All other industries | 32·3 |
| | 100·0 |

(including automobiles and lorries) were added to that used to produce all other machinery, the total would be about equal to that for group (6) above, something less than 5% of the industrial total.

Another way of subdividing energy use is by the desired end-form of useful energy. By this method, US industrial energy use in 1960 could be set out as follows:

**Table 2.6**    US industrial energy use in 1960

| End-form | Share (%) |
|---|---|
| Process steam | 44 |
| Direct heat | 31 |
| Work (mainly electric drive) | 21 |
| Electrolysis | 3 |
| Lighting, air-conditioning, refrigeration | 1 |
| Total | 100 |

*Transport*

As a country industrializes, both the amount and proportion of its gross energy consumption devoted to transport tends to rise. Now, fuel and raw materials must be hauled to factories, while the manufactured products must be carried to markets and ports. Then, as cities expand, workers and businessmen must be transported between home and work place; the bicycle no longer suffices. Finally, the need and desire for speed in the conveyance of passengers and some goods adds substantially to transport-energy consumption.

Both the geographic framework of a nation and its stage of economic development bear strongly on the amount of energy it uses in transport as well as on how the transport is accomplished. A contributing influence is the nature of the energy resources available.

If population centres and natural resources are relatively far apart, as they have been in North America, the transport share of energy expenditure will be greater than if the internal markets and the raw materials are relatively near each other, as they have been in the United Kingdom and Germany. If the principal available source of transport energy is brown coal or lignite, as it has been in most of central Europe, the energy consumed in transport will be relatively greater—because of the low conversion efficiencies involved—than where the main energy source is hydroelectricity, as in Switzerland, or crude oil, as in Japan. In the early stages of industrialization, the transport share rises fast, because raw materials and workers must be moved not only that finished goods be produced but that new plants, factories, and railroads be built. Then, as the rate of additions to plant levels off and as the efficiency of transport rises, the transport share declines—unless, as has been the case in the United States, the economic surplus obtained from the use of cheap fossil fuel is so great that the average individual can afford an automobile and air journeys for pleasure, in which case the transport share may reverse its decline (in the United States this reversal occurred in 1966).

Transport-energy consumption in the United States is shown in Table 2.7.

**Table 2.7**  Transport-energy consumption in the United States

| Form of transport | Share (%) |
|---|---|
| Automobiles, motorcycles | 58 |
| Lorries | 15 |
| Aviation (civilian and military) | 13 |
| Railroads | 3 |
| Buses | 1 |
| Marine and other | 4 |
| Oil and gas pipelines | 6 |
| Total | 100 |

In western (OECD) Europe the motor-car's share of total energy consumed in transport is almost identical (Leach, 1972).

One of the phenomena common to less-developed countries as they attempt to improve their economies is a single-minded reliance on highways, lorries, and buses for transport, to the extent of allowing more efficient rail and water systems to deteriorate toward disuse (Brazil is a striking example).

*Commercial and residential sectors*
The commercial and residential sectors are difficult to distinguish in energy accounting; energy-delivery systems commonly serve both dwellings and commercial establishments without care being taken to keep records in such a manner as to make the task of the energy analyst easy. In industrial society, the large-apartment complex, with retail establishments embedded in it, is not uncommon; a single electricity distribution system and a single natural-gas (or coal) delivery system serve the entire complex. The difficulty of allocating energy input between commercial and residential space in a nation of apartment-dwellers is the reason why these two sectors often are found lumped together, and as well for the shadow of doubt that lurks behind any tabulation which separates them cleanly without weaseling footnotes. The pith of one such tabulation for the USA is given in Table 2.8.

**Table 2.8**   Energy input in US residential and commercial sectors (Stanford Research Institute, 1972)

| End use | Shares in residential sector (%) | Shares in commercial sector (%) | Shares in both (%) |
|---|---|---|---|
| Space heating | 57·5 | 53·7 | 56·0 |
| Water heating | 14·9 | 8·4 | 12·3 |
| Air conditioning | 3·7 | 14·3 | 7·9 |
| Refrigeration and freezing | 7·5 | 8·6 | 7·9 |
| Lighting | 3·6 | 7·7 | 5·2 |
| Electric drive | 2·5 | 5·4 | 3·7 |
| Cooking | 5·5 | 1·8 | 4·0 |
| Television and clothes-drying | 4·8 | 0·1 | 3·0 |
|  | 100·0 | 100·0 | 100·0 |

Some of the principal end-uses for energy in the commercial and residential sectors will now be discussed in greater detail.

*Space or comfort heating*
Space or comfort heating (and cooling) takes place mainly in the commercial and residential sectors of the economy. Industrial society is urban society. In a rural agricultural society, or in a society based on fishing or forestry (in addition to agriculture), the individual worker is warmed by his clothing and his exertion. The space in which he works is not heated, except by the sun. At home he shares with his wife and children the heated space, which may be only one room of several. In industrial society, it is different. The worker in a factory or in an office commonly works in artificially heated or cooled space, and at home all rooms of the house are heated or cooled, not just one or two. In addition his wife may work outside the home, and there may thus during the day be three spaces heated to accommodate these two persons, where one would have sufficed in agricultural society. Consequently, the per capita demand for space heating in industrial society is multiplied.

Comfort heating and cooling seem to be addictive. A social phenomenon of high-energy society is the tendency to use inanimate energy instead of clothing to heat and cool the person, and then to raise the temperature of cold-weather interior heat and to lower the temperature of hot-weather cooling so that there is a considerable difference between the two. When this addictive trend spreads from the home to business and commercial enclosures, as it has in the United States, much energy may be consumed for small physiological returns and no material benefit.

In pre-industrial society most inanimate energy was used for space-heating, only a small amount for cooking, and even less for process-heating, such as in metal-working, and in glass and ceramic manufacturing.

Perhaps the most dramatic change in energy-input shares during the development of an industrial society is the rapid decrease in the space-heating portion. There are two reasons for this. Not only do the industrial and transport sectors grow, using energy mainly for work and process heat, but the efficiency of energy use for space heating improves to a greater extent than do efficiencies in producing work and applying process heat; for the latter reason, the amount of energy consumed to produce a unit of useful heat decreases substantially. Today, less than 30% of the energy consumed in the United States goes for comfort heating. The downward trend may be reversed soon, if the adoption of electric-resistance heating continues its rapid pace. The annual growth rate for electric space-heating in US homes has been more than 20% in recent years, and it is estimated that more than 20% of all new homes are electrically heated.

About 55% of all US space-heating is in homes, 35% in commercial buildings, and about 10% in industrial structures.

*Cooking and water-heating*
Although cooking and water-heating are basic needs of all but the most primitive humans, in industrial society they account for only a small portion (in the United States less than 7%) of total energy consumption, and if modern society were as dirty as early industrial society, as ignorant of hygiene, and as short of clothes and utensils to be cleaned, there would be still less energy required. The ratio of three times as much energy used for water-heating as for cooking in the United States is interesting, especially in view of the very much greater efficiencies of water-heating systems as opposed to cooking systems. It is perhaps atypical of industrial nations and may reflect the success of soap advertising and the proliferation of automatic dishwashers and washing machines more than it does cultural level.

*Refrigeration and air-conditioning*
Refrigeration or freezing of fresh or raw foods is a necessity in modern industrial society, because most of the people live far from the farm or seacoast, and because they need the fresh foods for health. It is also a convenience, because frequent shopping trips are no longer required, and a considerable variety of foods can be available in the home for quick preparation. To a great extent in late industrial society, refrigeration has taken the place in food processing occupied by curing and drying in early industrial society, and has made, despite its greater cost, substantial inroads into the field once peacefully occupied by tinning and bottling.

On a *per capita* basis, energy consumed for household refrigeration in the United States increased 66% in the 1960–68 period; for clothes drying,

97%; for food freezers, 135%; for television, 91%; for air-conditioning, 190%; for dishwashers, 299%; for washing machines, 45%; for cooking, not at all (based on data compiled by Stanford Research Institute, 1972).

For the same period in the commercial sector, marked increases were shown in energy consumption for air conditioning (93%) and for lighting and electric drive (about six times for the two combined).

*Lighting and electric drive*
No one who has come recently, after dark in good weather, upon an ancient city such as Cuzco or Dubrovnik, to find it sparkling with electric lights and projecting its halo on the velvet night, can have failed to wonder, however briefly, at the almost incredible changes electric lighting has brought to the life of man.

The smoky dim whale-oil lamp was replaced by the wicked kerosene lamp only a bit more than 100 years ago, and it was less than a hundred years ago that the Welsbach gas mantle first made it possible to read and work after dark in a light strong enough for detail. Shortly thereafter the first central electric generating plant was built and by the early years of the present century the proliferating wonders of electric lighting had bred an international electric cult, one of whose members, Lenin, was to proclaim that the future of Russia rested upon soviet *and electric* power.

Because electric lighting is everywhere in industrial society, because it is so important in our lives, because it is used with blatant extravagance to advertise wares of dubious worth, and because the making of light is woefully inefficient in energetic terms (most of the electric energy is converted to heat, only a small amount to light), it is somewhat alarming to discover that lighting accounts for only 5·2% of the total energy consumed in the residential and commercial sectors of North American society. Of course, in terms of electricity, lighting accounts for almost 20% of that used in these two sectors. In other industrialized countries, less electricity is used for air-conditioning and refrigeration, and relatively more for lighting.

The growing use of electricity to drive motors, not only in industry where that use accounts for a fifth of all energy used, but in commerce (5·4%) and the home (2·5%), reflects modern society's insistent trend toward powered mechanization, for lifts, computers, cash registers, erasers, shavers, cleaners, and so on.

There is a considerable amount of energy used in transport for electric drive, although energy statistics, including those used in this chapter, do not indicate it. Most large 'diesel' locomotives are diesel-electric locomotives, their diesel engines being used to drive electric generators whose output in turn drives electric motors geared to the powered wheels of the locomotive. Some large off-road vehicles, such as lorries used in opencast mines, also are diesel-electric, commonly having one motor to power each wheel.

### Television and clothes-drying

If lighting has had a revolutionary impact on human life, so have (1) long-distance communication of sounds and images, and (2) labour-saving devices in the home. The first is well illustrated by television, the second by clothes dryers.

The automobile, as well as the organization of industrial society, has tended to take the family out of the home and to fragment it. Television is a centripetal force acting to pull the family back into the home. Restrictions on the consumption of energy for non-productive transport may be accepted much more readily by a society which can substitute television for a day in the country than by one which cannot. Television can be a labour-saving device; countless mothers use it for 'baby-sitting', a tie-down and entertainment mechanism that frees the mother for other tasks.

**Table 2.9**    Percentage energy used in average US household by systems and devices, based on 1968 and 1969 data (Stanford Research Institute, 1972)

| Share of total energy consumed (%)* | System or device | Share of electric energy consumed (%) |
|---|---|---|
| 56·9 | Heating system | 11·5 |
| 14·7 | Water heater | 15·5 |
| 5·9 | Refrigerator | 17·4 |
| 5·4 | Stove and oven | 6·6 |
| 3·7 | Air-conditioner | 10·8 |
| 3·5 | Lighting system | 10·5 |
| 3·0 | Television | 9·0 |
| 1·9 | Food freezer | 5·6 |
| 1·8 | Clothes dryer | 3·6 |
| 0·4 | Irons | 1·3 |
| 0·4 | Washing machine | 1·0 |
| 0·3 | Electric skillet | 1·0 |
| 0·3 | Dishwasher | 0·9 |
| 0·3 | Coffeemaker | 0·9 |
| 0·3 | Radios | 0·8 |
| 0·2 | Electric blankets | 0·6 |
| 0·2 | Fans | 0·5 |
| 0·2 | Portable heaters | 0·5 |
| 0·1 | Vacuum cleaner | 0·4 |
| 0·1 | Toaster | 0·3 |
| 0·4 | Other* | 1·3 |
| 100·0 | | 100·0 |

* Includes broiler, hot plate, clocks, humidifier (and dehumidifier), mixer, shaver, food disposer, hair dryer, etc.

Those other tasks more and more involve the operation and monitoring of powered machines such as refrigerators, freezers, vacuum cleaners, blenders and mixers, tin-openers, coffeemakers and electric skillets, dishwashers, washing machines, and clothes dryers. The amount of energy required by such an appliance depends upon the efficiency of the process involved and the amount of use made of the appliance. Approximate percentages of energy used in the average US home for heating, cooling, cooking, lighting, television, and various appliances are shown in Table 2.9.

The other tasks of the housewife may include employment outside the home. Such employment, and its social repercussions like equal pay for equal work and other salients of the women's-liberation campaign, have been facilitated by the increasing use of electricity in the home; this has shortened the time required for acquisition and preparation of food and for cleaning, and has decreased or eliminated the handling of fuel and the monitoring of its combustion.

**Where it ends up**

*Time's arrow*
Unlike matter, energy cannot be recycled. Neither can it be diminished. This seeming paradox is resolved by the concept of entropy. Entropy has been defined in many ways, both qualitative and quantitative, as the tendency for all energy to degrade to heat at ambient temperatures; and as a measure of the heat in a system that cannot, because of degradation, be used again in that system.

However defined, entropy expresses reality. The degradation of energy, which Eddington called 'time's arrow', is unidirectional and irrevocable. Man speeds the natural increase in entropy by burning fuels at much faster rates than they would otherwise decay or oxidize. It is the quality of energy, its ability to do work and give useful heat, not its amount, that is subject to conservation. Energy conservation thus consists entirely in slowing the pace of man-induced entropy increase to some minimum consistent with mankind's present needs and desires, and with the existing level of concern for posterity.

Energy may be conserved in two ways: (1) by not using it, and (2) by using it in the most efficient way available. By almost any definition of need, man uses some energy needlessly. To some, the use of energy for growing, malting, fermenting, and distilling grain to make whisky is needless, if not downright sinful. To others, the use of energy to make heavy motor-cars capable of accelerating in a few seconds to a speed of 100 mph is unnecessary. And to still others, the use of high-grade energy (electricity) to advertise nostrums and notions is wasteful. To conserve by not using energy for such purposes probably would require the abandonment of consensual democracy, as well as the placing of severe constraints upon the growth-oriented market economy that encourages the expenditure of energy in most of the industrialized world.

From an environmental point of view, the first path (non-use), although strewn with thorns, may seem the better; but as long as we have a market economy among nations—whatever the form of the economy within nations—the quest for economic advantage and the fear of unemployment will provide strong pressures for production at high levels.

The second path of energy conservation is much more consonant with the principles of persistent production, to use energy as efficiently as possible without examination of the intrinsic worth of the aims of that use, and without consideration of the consequences of glut. The following section emphasizes certain aspects of the concept of efficiency as applied to energy systems.

## Efficiencies of energy use

### The Second-Law limit

Sadi Carnot, in one of the most elegant scientific papers ever written (1824), long ago explained why heat engines could never be highly efficient in converting the energy in their fuel into work. An efficiency of 100% could be attained, he demonstrated, only if the operating medium (steam or gas) could be exhausted at absolute zero ($-273°C$). This theoretical limit on the efficiency of a heat engine often is called the Second-Law limit, from the Second Law of Thermodynamics. The steam locomotives and steam pumps of his day illustrated Carnot's conclusion rather too well; they had a machine efficiency less than 1%.

Although great technological advances have been made, the Second Law has not been repealed. All power plants, fixed or mobile,* that derive their energy from combustion or fission of their fuel, may be regarded as limited in thermal efficiency by the Second Law of Thermodynamics; it tells us that, once we have degraded energy, we can upgrade it again only by the expenditure of more energy than we could retrieve from it, that reducing entropy in a system is a futile process in energy economy (it also tells us that perpetual motion is not possible).

As power plants were improved through better combustion, stronger materials, introduction of preheaters and turbines, lowering of condenser pressures and temperatures, their conversion efficiencies were raised to the levels shown in Table 2.10.

* Although the internal-combustion engine technically is not a heat engine, from an efficiency point of view it might as well be one.

**Table 2.10**   Conversion efficiencies of power plants

| | |
|---|---|
| Fossil-fuelled thermal-electric power-plant | 40% |
| Boiling-water reactor (nuclear) power-plant | 34% |
| Gasoline-powered internal-combustion engine | 25% |
| Diesel engine (large) | 38% |
| Gasoline-powered aircraft engine | 30% |

No further great improvements are expected with any considerable degree of confidence. Both the fossil-fuelled power-plant and the nuclear-fuelled power-plant have a theoretical maximum efficiency of about 60%; this will be costly to achieve, however, and may not be economic until the fuels become scarce.

Of the 64% of the US energy consumption that ends up as inutile heat, about half represents Second-Law 'waste'. In any discussion of energy conservation, Second-Law 'waste' should be considered, apart from waste heat generated in processes which can be, at least in theory, 100% efficient. Glass furnaces and blast furnaces, for examples, are thermally inefficient, but their inefficiency is a matter of design and process; it does not reflect an inherent limit less than 100 as do the inefficiencies of power plants.

*Electricity generating stations*
Fixed thermal power-plants rarely are able to convert more than 40% of the chemical or nuclear energy in their fuel into electricity; consequently, 60% or more of the energy in the fuel must be released at the plant in the form of heat. Not only may such release create an environmental problem, but it represents a great loss of a non-renewable resource.

Because of the cleanliness of electricity, and the ease with which it can be applied to motors in a wide range of sizes, to heating, cooling, freezing, and providing power for telephonic, radio, and television communications, the demand for it has been great. Only the cost of generating electricity and of the distributing networks and appliances has kept it from replacing direct heat and muscle power completely. Over the years the cost has been lowered by sustained improvements in the efficiency of generation and transmission.

At the start of the present century, thermal power-plants in the United States converted 3% to 4% of the chemical energy in their fuel into electricity, and much of the electricity was lost in inefficient low-voltage transmission lines. Today power-plant efficiencies can reach 40% and line losses have been reduced dramatically through the economies of scale of ultra-high-voltage transmission. The annual growth rate for electricity consumption in the United States for many years has exceeded the growth rate of total energy consumption. At the present time almost 30% of the total energy consumption in the United States is for the generation of electricity, but almost 40% of the inutile or waste heat produced by the national energy system comes from electricity generation and transmissions.

Questions of relative efficiency have been raised in regard to electrical heating as compared to direct heating with a fossil fuel. The facts are rather simple but the related judgments may not be. The pivotal point may be how the electricity is generated.

The system efficiency of space heating with electricity generated by fossil fuels is about 32%; all the loss is in the power-plant and transmission lines, since the heating process itself is 100% efficient or nearly so. On the other hand, if that fossil fuel is burned in a well-operated and designed furnace system, the comparable efficiency will be 75%. In this case, from the points of view both of conservation and of cost, the nod would go to the direct use over electricity.

Now, however, consider that the electricity is produced in a hydro-electric power-plant approximately 90% efficient in converting the kinetic energy of falling water into electricity. If the line loss is 8%, then system efficiency will be $90 \times 0.92 = 83\%$, somewhat better than direct heating with a fossil fuel. In addition, hydroelectricity is a resource with a depletion life considerably longer than that of the average fossil fuel. Moreover, it is clean and brings no problem of storage, stoking, or maintenance. The judgment, barring a substantial cost barrier, would be in favour of electrical heating.

Finally, suppose that the electricity is generated from nuclear energy which, like hydroelectricity, cannot be used for direct heating. In this case, the system efficiency today would be about $32 \times 0.92$ or 29%, contrasted to the 75% of coal, fuel oil, or natural gas used directly. Here the judgment may be based either on convenience (electricity) or conservation. Because present nuclear-power technology depends on a fuel base probably equivalent, in years of availability, to the longevity of petroleum and less than that of coal, the conservative judgment would go to the direct use of fossil fuel.

In most actual decisions below the national level, however, we are limited by existing systems for delivery and application of energy. In New York, for example, Corning Glass Works are replacing their gas-fired reverberatory glass furnaces with electric furnaces, not because electric furnaces offer greater thermal efficiency (they do) or greater system efficiency (they do not), but because natural gas within a few years will no longer be available, in the judgment of Corning managers, while electricity from nuclear fuel will be.

What all this comes down to is that electricity as a 'good' or 'bad' form of energy must be judged in the context of fuel availability as well as efficiency of delivery of useful heat and work. Only one generalization appears valid: electricity produced by the burning of a non-renewable fuel (fossil or nuclear) should not be used to perform a task which a fossil fuel, burned in another furnace, can perform with less draught on the remaining reserves of non-renewable fuel.

Consideration of system efficiency and fuel availability is 'of great importance in a world in which the natural fuels are so unevenly distributed that political control of them can make poor nations rich and rich nations poor.

System efficiency is likewise important in controlling the adverse environmental impacts of energy use. The high efficiency of burning fossil fuel directly for space heating may entail increased urban air pollution by fuels that contain sulphur and produce fly ash. The lower efficiency of electricity generation introduces more inutile heat into the environment, and may as well produce large quantities of sulphur oxides and fly ash, but this 'point source' can be located and designed to minimize the environmental impacts, whereas cities cannot.


*The motor-car and other machines*
Mobile power-plants and the machines they power, because they must operate without condensers and because of their accessory energy needs, are less efficient than large stationary power-plants. Early locomotives managed to convert less than 1% of the chemical energy in fuelwood or coal into useful work; present-day diesel-electric locomotives are much more efficient, having a machine efficiency of 25% or better, an almost threefold improvement over the average mid-century steam locomotive. This sort of technologic advance, which not only enables more work to be obtained from a despoiled unit of non-renewable natural fuel but also reduces the concomitant waste-heat discharge, has not occurred in the development of the automobile, a device that today is less efficient as a thermal machine than it was 50 years ago.

The modern automobile converts 5% or less of the chemical energy in its fuel tank into the work of propelling the vehicle (Thirring, 1968, p. 86). The remainder is exhausted and radiated from the engine or, after being converted to mechanical energy, is turned to heat by friction in the power train and by deformation of the tyres. The modern motor-car, at least in the United States, is twice as heavy as it needs to be for the tasks it is called upon to do. Because fuel consumption, at constant speed, is approximately proportional to weight, twice as much petrol is consumed, and twice as much waste heat produced as needs to be. Yet this blast furnace on wheels is fondly imagined to be an efficient and necessary transporting machine!

The automobile is inefficient because, at least in the United States:

(1) it is heavier than it needs to be;
(2) it is designed for comfort, to the extent that about 30% of the energy in its fuel is expended as heat from tyre deformation;
(3) it is used frequently for trips of less than five miles, during which it runs 'cold' and inefficiently;
(4) it has energy-consuming accessories such as radio and headlamps;
(5) it is designed for fast acceleration, and therefore is overdesigned for its ordinary use;
(6) it exhausts to the atmosphere at high exhaust temperature;
(7) during half its operation or more, it runs at less or more than optimum speed; and
(8) it normally carries far less than its optimum load.

The fuel used by US cars and small personal trucks is equivalent to the entire energy consumption of Japan, the world's third-largest energy consumer (Pierce, 1975). The economic difficulties recently imposed upon the United States by the oil-producing countries were made possible by the extravagance of the American motor-car.

Although other nations are not yet so wedded to the automobile, it appears that increases in efficiencies (thermal, machine, and system) in the use of this machine can be made in the near future, under the pressure of high energy costs and concern for national security. It is an expressed national strategy of the United States to improve the machine efficiency of the motor-car by 40% in the period 1974–80.

In energy terms the end-product of the energy flow through the motor-car is work. In engineering terms we say it is the work of propelling the vehicle and its occupants, in other words, useful work. The same is true of all machines. A machine is a device to convert energy into work. In this limited sense, even a human being may be regarded as a machine, one with a short-burst efficiency of about 30%.

In order to produce work, machines must have a power source. In classical antiquity, the power source on land was human and draft-animal muscle-power, and at sea was wind and slave-power. In the Middle Ages, the kinetic energy of falling water was harnessed by means of the water-wheel. In the High Middle Ages and the Renaissance, the kinetic energy of wind was caught by the sails of windmills, as well as of ships. The overwhelming power source of the Industrial Revolution was the coal-fired steam engine. Today the electric motor and the internal-combustion engine are the main power sources for machines.

As might be expected, the efficiency of a machine depends to a large degree on the nature of its power source. The most efficient machine in widespread use in modern society is the water turbine; impelled by falling water it can deliver as much as 98% of the kinetic energy of the water as mechanical energy or work. The least efficient machine (the motor-car) already has been discussed. Machines in whose motors or power sources the Second-Law limit comes into play are as a class inherently of much lower efficiency than those which can convert kinetic energy or chemical energy without combustion. Because machines of lower efficiency are needed in greater numbers or larger sizes to do a given amount of work, they impact their environment more than do machines of higher efficiency. It must be remembered that work is not a form of energy, and that in doing work, mechanical, kinetic, and electric energy are converted into heat. The efficiency of a given machine, therefore, cannot affect the rate of heat release into the environment because of its operation, but it may affect the distribution of that release. Electric-powered machines, for example, may entail release of much heat at a distant power plant, and only a small amount in the environment of use, while a gasoline-powered machine of the same size would release the total amount at the place of use.

*Space-heating and cooling*

Although we can calculate or estimate, from sales of fuel and electricity to households and apartment houses, how much energy is used in space-heating and cooling, trustworthy calculations of the overall efficiency of this energy use are not available. The operating efficiencies of a fossil-fuelled heating system can range from 25% to 75%, depending on its design, on the quality of its maintenance and operation, on the quality of construction of the structures whose interior space is being heated, and on the use pattern of that space. Available thermal-efficiency information relates in the main to new systems in new structures, and therefore represents ideal rather than contemporary practice. In the United States the overall efficiency of space heating probably is about 50%. It needs to be remembered that this is an engineering efficiency, in which there is an implicit assumption that all the space being heated needs to be heated, and to the temperatures maintained; as with the motor-car, the *social efficiency* of space-heating is much lower than its system efficiency.

The distinction of social efficiency from system or engineering efficiency can be illustrated by air-conditioning. Electric air-conditioning systems are about 50% efficient, again depending on the same variables as does the efficiency of space-heating systems. The social efficiency, on the other hand, relates to the degree that benefits of health, comfort, and production are achieved through the artificial cooling of living and working spaces. In hot humid climates there can be little doubt that a properly designed and operated air-conditioning system can provide comfort, may improve working ability (especially that requiring mental effort and manual dexterity), and may assist in maintaining good health. On the other hand, a system—like ice in drinks—used mainly to give a pleasant sensation may lead to unhealthy alternations of ambient temperature and humidity while yielding no improvement in work ability. The social efficiency of the latter system is low to negative.

During the past 15 years, the consumption of energy for comfort cooling in the United States has grown at the rate of about 20% per year. In Texas, a state with rather warm and long summers, electricity generating capacity is geared to summertime peak loads as much as 100% greater than the normal winter load; most of the electricity used in summer is used for air-conditioning; where 30 years ago there were only relatively few 'window' air-conditioners, today many homes and factories, as well as most office buildings, schools, churches, retail structures, and motor-cars, are centrally air-conditioned.

*Process heating*
The thermal efficiency of process heating may range even more widely than that of space heating. It has been estimated that some operating blast furnaces have an efficiency of only 5%. Reverberatory glass-melting furnaces have a thermal efficiency of 20%–35%. Overall in US industry the thermal efficiency of process heating probably is not greater than 25%; much improvement is possible, because there is no Second-Law limit to the efficiency of heating; the aim of efficiency here is to put as much heat into the material to be influenced as possible, and as little into the furnace or oven and its environment.

**Waste-heat concentrations**

*Power-plant effluents*
In industrial society much waste heat is produced by power plants, both the large fixed ones which generate electricity, and the small mobile ones which propel automobiles and other vehicles.

The heated effluent from large thermal power-plants, when discharged into bays, estuaries, natural lakes, or rivers, can cause changes that disrupt the pre-existing biotic community. The severity of the impact in terms of the life-support system, however, is a matter of controversy. One of the adjustments now becoming common is to create a man-made lake in which the rejected cooling water can cool to ambient temperature and then be re-used in the plant. An alternative is to instal cooling towers, which dissipate the heat into the atmosphere instead of into a water body. Wet cooling towers cool by evaporation of the cooling water; they may cause increased local fogginess. Dry cooling towers cool by radiation and convection; the cooled water is then recycled.

Increases in the scale of electricity generating plants and improvements in the efficiency of electricity transmission are opening new alternatives for dealing with waste heat. Power plants are being designed which will be mounted on floating offshore islands and will discharge their waste heat into the vast thermal sink of the world ocean. Also discussed is the possibility of forming clusters of power plants in subarctic locations, where waste heat might be used to keep bays from freezing over in winter or to support agriculture in a climate hostile to it. Except within the Soviet Union, such clustered power plants and their transmission grids would require carefully worked-out international arrangements for financing, power allocation, radioactive-waste handling, and security, because the large power-consuming nations are not those on whose territory the plants would be built.

*Urban heat springs*
Urban heat 'islands' or 'springs' are formed by heat emitted into the air from power plants, fixed and mobile; from heated buildings in winter and from air-conditioners in summer; from powered machines and appliances; from process furnaces and ovens; and from the bodies of humans and their warm-blooded pets and pests. The term 'heat island' is a static concept for what is actually a heat spring rising above a city sessile on the bottom of the atmospheric ocean. The spring forms and grows, not only with the growth of the city, but with the city's increasing use of inanimate energy for heating, cooling, processing, and work. The spring might diminish if such activities abated, the city's population decreased, or the efficiency of energy use in the city improved. Some urban heat springs have bloomed in recent decades to such an extent that they have changed local climates and may threaten man's life-support system if they expand and coalesce in relatively high latitudes; if waste-heat production continues to grow as it has, in a century or so the surface temperature of the earth may be a few degrees higher,* and there could be some spectacular consequences such as melting of ice-caps, a rise in sea-level, and submergence of all the world's coastal cities.

Of more immediate concern are the local climatic effects: increased average temperatures in the city, and increased cloudiness and precipitation (especially if the thermal effluent is accompanied by particulate emission). Although in winter the increased temperature may be a benefit,

---

* Although there is a puzzling secular cooling trend at present that may pose a different and more immediate threat.

in summer it can be a serious handicap both to comfort and to efficient use of energy resources, for air-conditioners set in motion to repel the advancing temperature reject heat into the atmosphere greater than that they extract from the air inside buildings, thus undermining their own effort and creating an ever-greater need for mechanical cooling.

### Radioactive waste energy

Radioactive waste materials, like other industrial wastes, are in gaseous, liquid, and solid forms. Unlike other wastes, however, they emit energy in ways potentially harmful to living organisms. The essence of radioactive waste is the energy it contains. Because a rapid increase in the production of electricity from nuclear power-plants is viewed by most industrialized countries as the only available alternative to increasingly scarce and inter-dictable fossil-fuel supplies, the problems of managing energetic wastes created during nuclear-power production are of increasing concern.

Under normal operating conditions it is possible to contain most, if not all, radioactive waste energy at any point in the fuel cycle except the starting point. The starting point of the nuclear fuel cycle is a mine; a mine, by definition, is within the biosphere; the wasting or decay of natural radioactive ores can be neither stopped nor delayed. Fortunately, natural ores yield decay products of relatively low radioactivity, and the welfare of the miners can be protected by proper ventilation and by exposure constraints. Waste material from milling of the ore can be isolated from human activities; although much more bulky than the strongly radioactive wastes created by fission, the mill 'tailings', like the ore, have a low level of radioactivity.

Metallic uranium and thorium may be handled with little or no risk; consequently, the energetic waste from fabrication of fuel elements has little hazard potential—unless those elements are to contain plutonium, an extremely hazardous material.

It is fission in a reactor which produces what has been called the *ultimate pollutant*—highly-radioactive long-lived isotopes that must be kept out of the biological environment for hundreds or thousands of years.

Although the bulk and weight of all high-level radioactive waste expected to be produced during the next 50 years are minor, the amount of radioactivity and the potential for harm in that waste will be far from minor. Because of the long-hazard lives of many of the radio-isotopes, releases to the environment are cumulative as well as irretrievable. Because some radio-isotopes tend to be concentrated biologically in the food chain, releases believed to be harmless may become hazardous. Because ionizing radiation is both carcinogenic and mutagenic, it is possible that releases representing little or no somatic risk to present generations may shorten the lives of the younger of those generations and may be catastrophic to a future generation. For these reasons, the proper management of radio-active wastes poses one of the most serious challenges to the ability of mankind to protect itself from its own mistakes.

## FURTHER READING

Carnot, Sadi (1943) [1824], *Reflections on the Motive Power of Heat and on the Machines fitted to develop this Power* (translated by R. H. Thurston), New York, Amer. Soc. Mech. Engineers.

Cook, Earl (1971), 'The Flow of Energy in an Industrial Society', *Scientific American*, v. 224, pp. 134–144.

Cook, Earl (1972), 'Energy for Millennium Three', *Technology Review*, v. 15, n. 2, pp. 1–8.

Cook, Earl (1975), 'Ionizing Radiation', chapter 13 in *Environment* (2nd ed.), Stamford (Conn.), Sinauer.

Council on Environmental Quality (US) (1972), *Environmental Quality* (3rd annual report), Washington, Govt. Printing Office, 450 pp.

Evelyn, John (1933) [1661], *Fumifugium: or the Inconvenience of the Aer and Smoake of London dissipated* (reprint), London, Oxford University Press, 49 pp.

Leach, Gerald (1972), *The Motor Car and Natural Resources*, Paris, OECD, 54 pp.

Pierce, J. R. (1975), 'The Fuel Consumption of Automobiles', *Scientific American*, January, pp. 34–42.

Stanford Research Institute (1972), *Patterns of Energy Consumption in the United States*, Washington, DC, Office of Science and Technology, Exec. Off. President, 156 pp.+ appendices.

Thirring, Hans (1968), *Energy for Man*, New York, Greenwood, 409 pp.

United Nations Department of Economic and Social Affairs (1973), *World Energy Supplies 1968–1971*, Statistical Papers Series J, n. 16, 187 pp.

CHAPTER THREE

# ECONOMICAL USE OF ENERGY AND MATERIALS

Andrew Porteous

*One day, perhaps, it will begin to dawn on our masters
that to plunder, in a single generation, resources that
have taken 350 million years to accumulate, is a crime
for which our descendants will never forgive us.*

Lord Avebury, *New Scientist*, 22 Feb., 1973

The title of this chapter suggests a treatment of the laws of thermodynamics, energy conversion and utilization, and the various means for
increasing conversion efficiency. While energy is a fundamental resource,
it is not divorced from other resources necessary for modern living. We
need raw materials to make copper, steel, plastics, fertilizers, etc. These
materials must be processed, e.g. by mining, crushing, flocculation, smelting, forging, fabrication, transport. At each stage energy is consumed, but
if the available ore concentration for, say, copper, were to be reduced
overnight by 50%, then the energy consumption to obtain the same
amount of copper as before would be at least doubled, the 'pollution'—
or impact the operation has on the environment—from this operation
would also be at least doubled. Thus energy conservation, to give it its
popular title (*energy economy* would be much better) is tied to resource
conservation in general. This chapter, then, is on the economical utilization
of both energy and materials.

## Natural cycles and the Laws of Thermodynamics

The need for energy is everywhere—a statement which may be illustrated
by the following:

63

(a) It is impossible to add to the material resources of the world.

Resources fall into two categories, renewable and non-renewable. Timber and water are examples of renewable resources; they are, of course, finite in their availability. Fossil fuels, copper, iron ore, are examples of non-renewable resources. The time scale is the determining factor; that for a timber crop is roughly 40 years, whereas oil reserves were formed over 350 million years.

(b) It is impracticable to dispose of waste materials outside the world and its envelope of air—excluding the projected frivolity of sending radioactive wastes to the sun.

Because of the implications of statements (a) and (b), nature has evolved a complex series of interrelationships whereby the essential materials required for life support are recycled and returned in a re-usable form. As an example, carbon is a principal element in the formation of natural compounds and, as it is in finite supply, a natural renewal cycle (the carbon cycle) has evolved so that it is made available for re-use. The atmosphere contains $700 \times 10^9$ tons of carbon in the form of carbon dioxide. Plant life draws in and consumes $35 \times 10^9$ tons per year carbon in the form of $CO_2$. There are also $450 \times 10^9$ tons stored in *living* tissues and cells, i.e. atmospheric $CO_2$ (at a rate of $35 \times 10^9$ tons per year) is fixed by plant life via photosynthesis. However, the respiration of the plants themselves releases $10 \times 10^9$ tons per year as $CO_2$ back to the environment. The living tissues die at a rate of $25 \times 10^9$ tons per year and enter the dead organic zone where at any one time $700 \times 10^9$ tons of carbon are stored. In the dead organic zone, the decomposer bacteria operate and respiration accounts for $25 \times 10^9$ tons per year from their myriad operations.

Now we have in the carbon cycle:

(1) A balanced system with (relatively) constant-capacity reservoirs.
(2) Several flow paths which together balance, i.e. with no loss or gain of mass.
(3) A continual recycling of material from atmosphere to living matter to dead matter and back to atmosphere, i.e. material is restored to its original state by a cyclic process.
(4) An external supply of energy (from the sun) to power the cycle.

We shall return to these points later. Note, however, that the carbon cycle components are renewable resources, such as trees and grass, and that the cycle is powered by an external power source, solar energy. In fact the carbon cycle obeys the laws of thermodynamics which apply to all known phenomena.

*The First Law of Thermodynamics*
This law is a statement of the conservation of energy, i.e. energy can neither be created nor destroyed. Energy has many forms—electrical, chemical (the combustion of coal is a conversion of chemical energy), thermal, nuclear, etc. The First Law tells us that the sum of energy in all its various forms is a constant. Physical processes change only the distribution of

energy, never the sum. However, the First Law tells us only that energy is conserved. It does not tell us the direction in which processes operate; this is the province of the Second Law of Thermodynamics.

## The Second Law of Thermodynamics

This Law specifies the direction taken by physical processes. Its usual statements follow the lines that heat transfer takes place from a hot body to a cooler body, or that 'it is impossible to construct a system which will operate in a cycle, extract energy from a reservoir and do an equivalent amount of work on the surroundings'; in other words, a heat engine, converting heat to some other form of energy, cannot operate at 100% efficiency.

Thus the Second Law tells us, for instance, that concentrations of high-availability materials or energy will, in general, disappear, e.g. a sugar lump will dissolve in water, high-availability energy will degrade to low-availability in use, or in general that order becomes disorder. The general statement of the Second Law is that everything proceeds to a state of maximum disorder, and by implication is less useful in the event of the disorder occurring, e.g. a sugar lump is more useful when it is not dissolved in water. A kettle of boiling water is more useful than the same amount of water mixed in a bath of cool water. All these examples are manifestations of the Second Law, which applies to all biological and technological processes and cannot be circumvented.

The relevance of the Second Law for us is that the progression from order to disorder demands the inevitable degradation of energy from the usable to the unusable. While the First Law states that the total amount of energy in the universe is constant, the Second Law states that the fraction of energy available for use is constantly diminishing, i.e. that the universe is running down and that man is contributing to this process. We consume ordered materials such as petroleum, coal and sugar to provide energy for our way of life. The end result is degradation of energy at high temperature to the same amount of energy at low temperature—in a state in which it is much less useful.

This degradation of energy ultimately manifests itself in the production of thermal energy (heat) at relatively low and thus useless temperatures. The heat emission from a car-engine exhaust, the hot-water discharge from a power station, the heat of tyre friction on the road, the heat loss from the body, the heat of a decaying carcass are all forms of relatively unusable and degraded energy.

The laws of thermodynamics are often used to determine the ideal efficiency for a given process and to explain why 100% conversion efficiency cannot be achieved in any process concerning thermal energy.

They govern the behaviour of all living systems. They tell us why we need a steady input of energy to maintain ourselves; why for a cow to gain 1 kg it must eat much more than 1 kg in food; why to generate 1 kW of electricity the thermal power input required is much greater than 1 kW in fuel.

They impose limitations on how far the earth will support species: if it can grow only a certain amount of food (energy), then the laws impose restrictions on the number of mouths which can be fed.

There are essentially two categories of life:

(1) The producers, which transform solar energy and $CO_2$ into sugars via photosynthesis, i.e. plants.
(2) The consumers, i.e. animals who live directly or indirectly off the producers.

Thus we may picture the flow of energy through this system as steps in a chain. A chain starts with green plants which are the producers; they are the first trophic or feeding level. At the second trophic level come the herbivores, the primary consumers. At the third trophic level come the first-order carnivores, and so on. Man's correct role is that of a herbivore, as he eats mainly vegetable matter; but he is also a secondary consumer when he eats meat. At each transfer of energy, there is a loss. Approximately 10% of the energy stored in plants may turn up as available energy in the herbivores, and roughly 10% of their energy may be incorporated into the first-order carnivores. The Second Law of Thermodynamics cannot be circumvented, but a knowledge of its consequences can enable energy to be more efficiently utilized. As an example, consider that 1000 kg of wheat produce 100 kg of cattle, which in turn produce 10 kg of human tissue. By moving man one step down the food chain, ten times as much energy is directly available, i.e. 1000 kg of wheat can now produce 100 kg of human tissue. Similar, if less dramatic, results can be obtained in other energy utilization processes.

*Fossil-fuel energy conversion*
Consider the energy flow in a power generation system from combustion through to end use (figure 3.1). Let us put some numbers on this flow diagram so that we may scale the efficiency of the process. The fuel supplies 1000 kW input power and it is seen that the various losses incurred in the associated heat transfer and energy conversion processes total 700 kW. Thus our power generation system has an efficiency of 30%. (The most modern stations have efficiencies of 40%.) Now consider the fate of the electrical energy. If it is to be used for hot-water heating, it will supply 300 kW for this purpose, but if we move our water heater one stage down the system, we can have 850 kW input (in the form of heat) to our hot water and not 300 kW (as electricity) for the same input of fuel. This is

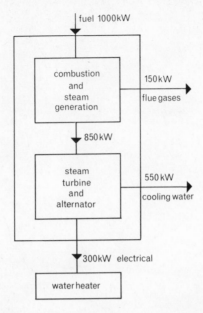

**Figure 3.1**   Energy flow in a power generation system.

analogous to man eating grains as opposed to beefsteaks. The important point here is that using thermal energy from combustion to generate electricity which in turn is used for heating is a very inefficient process; it incurs a much greater energy expenditure than would be obtained by bypassing the generation stage completely. We shall take up these points in detail when power generation and total energy concepts are discussed.

## Resources: energy and materials

*Resources are not, they become . . .*

At no time has this quotation been truer than the present day. The scale of resource consumption is what matters, and this is best illustrated with reference to crude oil. Figure 3.2 shows world crude-oil production from 1880–1972. The exponential nature of the growth curve is clearly evident, and illustrates the reasons behind the fears of all informed people that such rates of growth cannot be sustained.

An idea of the lifetime of oil reserves at present rates of consumption can be obtained by estimating, from geological data, the amount of oil that was initially present. If the cumulative production at any time is known, then the reserves are the difference between the initial supply and

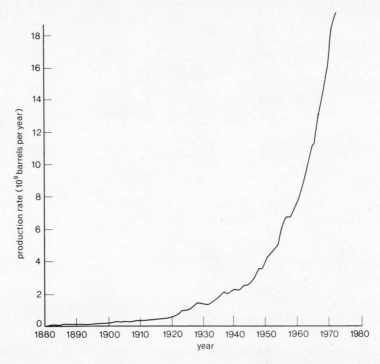

**Figure 3.2**   World production of crude oil, showing exponential growth (M. K. Hubbert, 1969 updated[1]).

the cumulative production. Eventually the reserves will decline to zero. The complete production cycle is characterized by an exponential rise at the beginning and an exponential decline towards the end. The total area under the curve is the ultimate amount of fuel $Q_\infty$ that can be produced during the cycle. It is salutary to predict the complete cycles of world crude oil production for the upper and lower limits of the estimated total oil reserves, i.e. $Q_\infty = 1350 \times 10^9$ barrels and $2100 \times 10^9$ barrels (King Hubbert, 1969), as in figure 3.3. For the smaller figure, a peak production rate of $25 \times 10^9$ barrels per year is estimated to occur around 1990, with the middle 80% of the cumulative production requiring only the 58-year period from 1961 to 2019. For the higher figure the peak of the production rate of about $37 \times 10^9$ barrels per year is delayed to year 2000. Thus a grave oil shortage can be foreseen by the turn of the century—in fact we already have such problems, albeit partly political.

The need for energy conservation should now be obvious. The oil is not

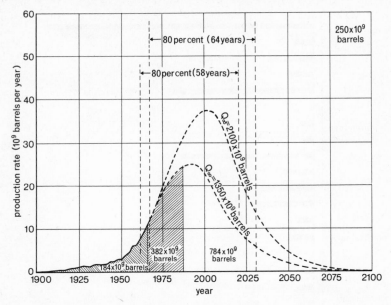

**Figure 3.3** Complete cycles of world crude-oil production for two values of $Q_\infty$ (M. K. Hubbert, 1969).

just an energy resource; it is a supply of large organic molecules, the raw materials of the chemical industry.

To complete the resource picture, figure 3.4 shows the lifetimes of estimated recoverable reserves of mineral resources. Clearly, coal is going to outlast oil by several centuries, but we can expect coal to be turned into liquid fuels by the turn of the century. It is sobering to observe that many of the materials currently taken for granted can be expected to be in short supply in the next 1–5 decades, unless economic forces and/or intelligent resource conservation impose a steady-state rate of energy consumption or zero growth; there is, as yet, no indication of either of these.

The argument is sure to be raised that as high-grade ores run out, low-grade ores will be processed—even ultimately from the sea. The sources of energy from fossil supplies clearly do not support these notions. Then what about fusion power? Future generations may have an everlasting energy source from nuclear fusion; we have not. Even if this dream becomes reality, it will not sustain us for ever. The Second Law of Thermodynamics dictates that all energy is ultimately degraded to heat (this neglects energy of deformation, chemical bonding, etc.) which is rejected within the earth's envelope and ultimately radiated to space. However, if this energy input to the earth were to approach 1% of the annual solar

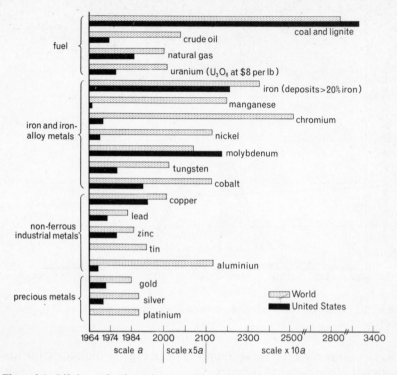

**Figure 3.4** Lifetimes of estimated recoverable reserves of mineral resources. Reserves are those that are of high enough grade to be mined with today's techniques. Increasing population and consumption rates, unknown deposits, and future use of presently submarginal ores are not considered (Cloud, 1968).[2]

input, we could expect major alterations in the world's climate.[3] This places an ultimate restriction of an energy input of about $10^{16}$ kWh/annum on the fusion power sources yet to be developed. Thus even in the future there will be restrictions placed on rates of energy use and, consequently, rates of materials extraction and processing. Conservation both of energy and materials is therefore fundamental if standards of living are to be maintained and resources are to be passed on for the use of future generations.

*Energy conversion efficiency and the concept of total energy*
To consider energy conversion efficiency, we must consider some of the forms that energy can take, and the usual conversions between these forms. The principal source of energy in the United Kingdom and industrialized nations derives from fossil fuels, as discussed in chapter two. Fossil fuels

contain chemical energy, which is usually converted by the process of combustion to thermal energy as in a coal fire. Often steam is raised in a boiler, and the thermal energy used to drive a turbine which converts the thermal energy to mechanical energy. Now the mechanical energy can be used to provide power, or can be converted in a generator to electrical energy.

There is a most important distinction between thermal and mechanical energy. Mechanical energy is 100% available, i.e. it is completely convertible to work; thermal energy is not completely available. The theoretical maximum efficiency of a heat engine is

$$\frac{T_1 - T_2}{T_1}$$

where $T_1$ is the absolute temperature (K) of the working fluid (e.g. steam) entering the device, and $T_2$ is the absolute temperature of the fluid leaving the device. This is known as the Carnot efficiency.

Absolute zero is the lowest possible temperature that can be achieved (absolute temperature = °C + 273). Thus if our thermal device could be operated in outer space, the exit or exhaust temperature $T_2$ could approach absolute zero, and the theoretical conversion of thermal to mechanical energy would approach 100%. But $T_2$, the exhaust temperature on planet earth, is restricted by the temperature of the surroundings and in practice must be greater than the surrounding temperature. In a typical heat engine $T_2$ is about 40°C or 313K, and $T_1$ is unlikely to exceed 800K. The maximum theoretical conversion that can be obtained is therefore

$$\frac{800 - 313}{800} \times 100 = 61\%$$

Now the Carnot efficiency was computed on the basis that all the high-grade thermal energy was introduced at a constant upper temperature of 800K (the current temperature limit of alloy steels). However, the operating cycle for steam turbines does not conform to Carnot's heat engine, as all the thermal energy is not so introduced. The corrected maximum theoretical conversion is 53%, and the state of the art of modern turbine technology is such that around 46% would be attained, i.e. 87% of what is theoretically attainable. To improve on this would be very difficult, and it is a great tribute to turbine designers that this has been obtained.

To obtain the overall efficiency of a steam power plant we require:

(1) the conversion efficiency of the boiler;
(2) the conversion efficiency of the turbine;
(3) the conversion efficiency of the generator.

The best modern boilers have an efficiency of about 85%, i.e. 85% of the chemical energy of the fuel is converted to thermal energy. The turbine efficiency will be taken as 46%, and the generator efficiency as 99%. The overall efficiency is the product of the component efficiencies, i.e.

$$(0.85 \times 0.46 \times 0.99) \times 100 = 39\%$$

which approximates to the efficiency of the most-modern fossil-fuelled power stations. Nuclear power stations use steam raised at a lower temperature than in fossil-fuelled plants, and their efficiency is currently about 30%. A nuclear station wastes more heat than a comparable fossil-fuelled station.

We are now in a position to analyse a typical heating situation.

*The case for total energy*

Three alternative schemes are worth considering.

Scheme 1 has energy input $E_1$ in the shape of natural gas. This is fed into a modern power station, where 40% is converted to electricity and 60% is rejected as waste heat. The electricity is used to provide hot water in 200,000 flats, all fitted with immersion heaters which are 100% efficient.

Scheme 2 uses natural gas directly for the 200,000 flats whose water heaters average 60% efficiency.

By comparing schemes 1 and 2, it can be seen that substitution of gas (a primary fuel) for electricity effects a substantial energy saving—in this case 50%, as $E_1 = 1.5E_2$ for the same heating load. (Note if the overall efficiency of scheme 1 were 30%, then $E_1 = 2E_2$.)

Now consider scheme 3 where steam is raised, power is generated and, instead of rejecting the heat from the turbine at the lowest possible temperature, it is rejected at a higher temperature, suitable for water heating. A back-pressure turbine gives exhaust steam at a temperature well above 100°C. The generating efficiency of this system is substantially lower than in scheme 1, but the overall thermal efficiency is (in theory) 100% as all the reject heat is now utilized, and electrical energy is also available for useful mechanical purposes.

For a heating load of 100 units, the energy requirements of the three schemes will be as follows:

| | |
|---|---|
| scheme 1 (all input used to generate electricity) | 250 |
| scheme 2 (all input used to heat water) | 168 |
| scheme 3 (electricity generated at reduced efficiency; waste used to produce hot water) | 142 |

This concept of scheme 3 is known as *total energy*. In practice there will be losses in flue gases, heat exchangers, etc., and the total energy input would probably be about the same for schemes 2 and 3, *but* electricity is generated in addition to servicing the same heat load as schemes 1 and 2.

Total energy is not new; it is used in many factories and process industries where the electrical and heating demands are suitably balanced; there must also be a distribution system for the exhaust steam or hot water. New York has an underground mains system supplied with steam by the Con Edison company. Total energy is not a panacea but, in areas where substantial high-density housing is being built, there could be a case for the construction of a small power station providing electricity and heating from the same installation.

A large modern power station (2000 MW electrical output) would, on the figures of scheme 3, produce 4000 MW as heat for district heating. Assuming a steady heat load of 4 kW per house, an outlet of a million homes would be required. However, as costs continue to rise, total energy may become more attractive.

District heating is one total-energy concept. Others that have been proposed are power generation and sea-water distillation, whereby combined power generation and sea-water distillation plants are run on base-load conditions, thus providing an alternative means of water supply for the United Kingdom, as opposed to the construction of new dams or estuarial barrages. However, these schemes have not been adopted, as it is claimed that conventional catchment methods are capable of supplying cheaper water than desalination.

An important concept arising from the foregoing section is that a distinction must be drawn between the various forms of energy—primarily between electrical and thermal energy. The former is 100% available; the latter is not. The subscripts $t$ and $e$ will be used to distinguish between the two forms.

Table 3.1 shows various energy conversion devices and their efficiencies. The energy conversion stages used are also shown. Fuel cells, whose efficiency is around 50%, require hydrogen and oxygen produced by electrolysis. Therefore the overall efficiency of a fuel-cell system is the product of the electrical-generation efficiency and the fuel-cell efficiency. Assuming 30% as a more accurate picture of the former efficiency, then a fuel-cell system has an overall efficiency of 15%. Advocates of the hydrogen economy (whereby hydrogen replaces fossil fuels as a power source) should note that the production of hydrogen involves electricity production which, in turn, means fuel consumption of one form or another; then, as noted previously, even with the advent of fusion the Second Law dictates that energy will be degraded to heat, which must be disposed of by re-radiation to space.

The consequences of the Second Law are unavoidable.

**Table 3.1** Energy conversion devices and their efficiencies (from 'The Man-Made World', Technology Foundation Course, The Open University)[(4)]

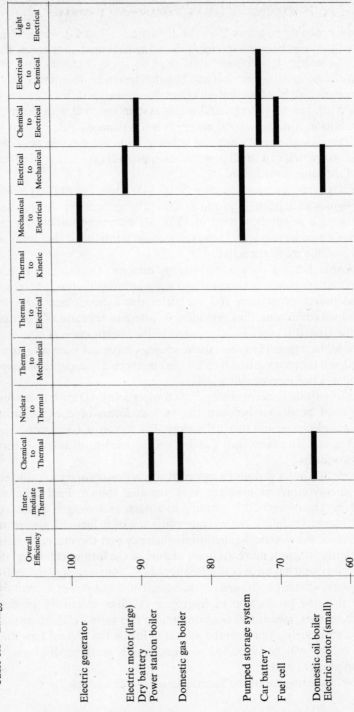

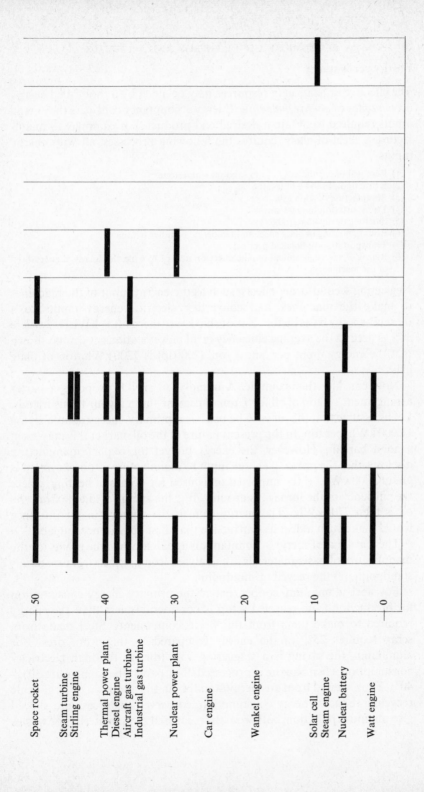

**Energy accountancy**

The effective utilization of resources may be aided by a tool called *energy accountancy* or *energy budgeting*. Energy accountancy considers the energy inputs required to obtain a desired end product. For example, to manufacture a sheet of glass requires the following processes, all with energy inputs.

(1) Removal of ground cover from sand pit + limestone.
(2) Extraction of sand + limestone.
(3) Manufacture of soda ash.
(4) Transportation to glass works.
(5) Smelting of raw materials.
(6) Rolling/floating of glass sheet and tempering or other treatment.
(7) Transportation of finished product.
(8) Repair of and replacement of machinery on items 1 to 6 (i.e. the energy 'depreciation' of the machinery).

Neglecting second-order effects, such as the energy input of the machines to make the machines, and converting electrical-energy inputs to a thermal-energy basis by the relationship kWh (thermal) = $kWh_e/0.3$ where 30% is used as the average efficiency of all power stations, it can be shown that the energy input per 'short ton' (2000 lb) is $7200\,kWh_t$/ton of plate glass.[5]

Now consider this example. A friendly oil sheikh proposes a barter arrangement—1 ton of oil for 1 ton of cement—as a favour to his friends. Let us assume that the oil is to be used as a primary-energy source giving $11,000\,kWh_t$ per ton. In the present nature of the oil market this may seem a good bargain. However, the energy budget for cement manufacture discloses that the primary-energy input to manufacture a ton of cement is about $5000\,kWh_t$. If the imported oil is used for domestic heating, where the efficiency of the furnace is around 70%, the energy available from the oil is then $7700\,kWh_t$. The energy gain to the country in this instance would make such a deal less attractive than first appearances suggest.

The real value of energy accountancy is apparent when the inputs for the manufacture of basic materials are analysed from (1) the virgin materials standpoint, (2) the recycling standpoint.

For steel, aluminium, copper, paper, the primary-energy consumption for production from recycled scrap is considerably less than the energy required to make them from the virgin components. Steel made from scrap requires 25% of the energy compared to that from ores. For aluminium, the saving is a staggering 95% (due to the high electricity consumption when bauxite is processed). For paper, the savings are 30%–40%. Thus, viewed from an energy standpoint, the adoption of large-scale recycling reduces energy consumption, conserves resources, and should also minimize pollution, both from the recovered scrap materials or wastes themselves and the dereliction, water pollution, grit, fumes and noise generated in the extraction processes. The recycling of domestic refuse is considered in detail later. The potential benefits of recycling in general are highlighted here.

Table 3.2 gives the energy inputs for basic materials processing from raw ores. The basis is kWh (thermal) per short ton (2000 lb).

**Table 3.2** Energy consumption in basic materials processing* (*Environment*, June 1972)

| Material | Energy for unit production (kWh$_t$/ton) | Machinery deprecia- tion (kWh$_t$/ton) | Transpor- tation (kWh$_t$/ton) | Total (kWh$_t$/ton) | No. of tons consumed (1968) | Total energy (kWh$_t$) | % of total energy consumption ·(1968) |
|---|---|---|---|---|---|---|---|
| 1. Steel (rolled) | 11,700 | 700 | 200 | 12,600 | $90 \times 10^6$ (excluding alloys) | $1·13 \times 10^{12}$ | 6·20 |
| 2. Aluminum (rolled) | 66,000 | 1,000 | 200 | 67,200 | $4·07 \times 10^6$ | $2·74 \times 10^{11}$ | 1·49 |
| 3. Copper (rolled or hard drawn) | 20,000 | 800 | 200 | 21,000 | $2 \times 10^6$ | $4·2 \times 10^{10}$ | 0·23 |
| 4. Silicone, metal, and highgrade steel alloys | 58,000 | 1,000 | 200 | 59,200 | $2 \times 10^6$ | $1·19 \times 10^{11}$ | 0·65 |
| 5. Zinc | 13,800 | 700 | 200 | 14,700 | $1·5 \times 10^6$ | $2·2 \times 10^{10}$ | 0·12 |
| 6. Lead | 12,000 | 700 | 200 | 12,900 | $0·467 \times 10^6$ | $6·05 \times 10^9$ | 0·04 |
| 7. Misc. electrically processed metals | 50,000 | 1,000 | 200 | 51,200 | $2 \times 10^6$ | $1·02 \times 10^{10}$ | 0·06 |
| 8. Titanium (rolled) | 140,000 | 1,000 | 200 | 141,200 | $16 \times 10^6$ | $2·24 \times 10^9$ | 0·01 |
| 9. Cement | 2,200 | 50 | 50 | 2,300 | $74 \times 10^6$ | $1·7 \times 10^{11}$ | 0·93 |
| 10. Sand and gravel | 18 | 1 | 2 (short- distance hauling) | 21 | $918 \times 10^6$ | $1·83 \times 10^{10}$ | 0·10 |
| 11. Inorganic chemicals | 2,400 | 100 | 200 | 2,700 | $67 \times 10^6$ | $1·8 \times 10^{11}$ | 0·98 |
| 12. Glass (plate finished) | 6,700 | 300 | 200 | 7,200 | $10^7$ | $7·2 \times 10^{10}$ | 0·39 |
| 13. Plastics | 2,400 | 300 | 200 | 2,900 | $6 \times 10^6$ | $1·74 \times 10^{10}$ | 0·10 |
| 14. Paper | 5,900 | 300 | 200 | 6,400 | $50·7 \times 10^6$ | $3·24 \times 10^{11}$ | 1·77 |
| 15. Lumber | 1·47 per board ft. | 0·02 per board ft. | 0·02 per board ft. | 1·51 per board ft. | $3·75 \times 10^{10}$ board ft. | $5·66 \times 10^{10}$ | 0·31 |
| 16. Coal | 40 | 2 | — | 42 | $556·7 \times 10^6$ | $3·3 \times 10^{10}$ | 0·20 |

\* Accuracy ±20%

Table 3.3 gives a selection of energy contents of materials and manufactured products. The contribution of energy costs to product value is quite significant in many cases, and is an illuminating comment on our energy-intensive society.

*Housing*

An energy input analysis for a 3-bedroomed semi-detached house, constructed to Parker-Morris standards, is shown in Table 3.4.

The figures speak for themselves. Consider, for example, the alternatives of soil cement blocks or rammed earth; the difference in energy inputs is striking.

A very worthwhile exercise would be the analysis of the total-energy requirements of basic materials in relation to their total life. Another aspect

**Table 3.3**   Typical energy contents of materials and manufactured products (NATO, 1974)[6]

|  | Energy* megajoules/kg | Cost of energy*/ Value of product |
|---|---|---|
| *Metals* | | |
| Steel (various forms) | 25–50 | 0·3 |
| Aluminum (various forms) | 60–270 | 0·4 |
| Copper | 25–30 | 0·05 |
| Magnesium | 80–100 | 0·1 |
| | | |
| *Other products* | | |
| Glass (bottles) | 30–50 | 0·3 |
| Plastic | 10 | 0·04 |
| Paper | 25 | 0·3 |
| Inorganic chemicals (average value) | 12 | 0·2 |
| Cement | 9 | 0·5 |
| Lumber | 4 | 0·1 |

* These are typical values. The actual value depends on the purity, form, manufacturing process and other variables.

is that of running costs. The heating costs of homes and industrial buildings can be substantially reduced by the adoption of thermal insulation and the use of proper materials. Thus cheap construction, using claddings of glass on concrete or steel, may in the long run be very costly in terms of the energy performance of the building. However, other factors militate against the adoption of efficient energy utilization practice.

Consider the case of air-conditioned office buildings as outlined in the following extract:[7]

Current air-conditioning systems for office buildings have an inbuilt potential for wasting fuel and power. Buildings with air conditioning frequently use 4 or 5 times as much electrical power as those with heating and mechanical ventilation. The capital cost of the most expensive air-conditioning system is 2 or 3 times that of the least expensive, and is greatly influenced by the building design.

The electrical operating costs are usually predicted on the basis of annual hours of full-load operation. It is current practice to assume that refrigerating plant, for example, operates for over 1000 hours per annum at full load. In some buildings the actual period of full-load operation is twice this (2000 hours), a fact illustrated by observing how many air-conditioning refrigerating plants have to be brought into operation even when the outside air temperature is 40°F or less. The theoretical period for full-load operation for an average South-of-England summer is between 500 and 900 hours, the latter figure being for a 24-hour day, 7 days a week. Equivalent full-load operation for the normal office building is about 70 hours per annum. There are, therefore, potential power savings on refrigeration alone of between 30% and 70%, depending upon the quality of the design and of the control system. Other sources of waste are the fans and pumps, which frequently run twice as long as is really necessary, judged by the length of the normal working year. The fans of high-velocity air systems frequently use more power than the refrigeration.

The scope for saving considerable amounts of energy clearly exists: why, then, is energy conservation not practised? The answer lies in the lack of financial incentive to do so. Air-conditioned buildings in London have an

**Table 3.4** Energy input analysis for construction of 3-bedroomed semi-detached house (McCillop, 1972).[8] Materials and energy requirements.

| Materials | Energy Inputs | Site Preparation |
|---|---|---|
| Bricks: 16,000 | 3,200 kWh | Excavation/Handling: |
| Steel: 1·2 tons | 9,200 kWh | 2,000 cu ft = 6,000 kWh |
| Glass: 320 ft$^2$ | 2,000 kWh | Cement mixing and Miscellaneous machinery: |
| Concrete: 10 tons | 5,000 kWh | 100 gal fuel = 4,200 kWh |
| Cement: 2 tons | 3,600 kWh | Subtotal 10,200 kWh |
| Plaster: 3 tons | 900 kWh | |
| Timber: 4·3 cu m | 310 kWh | |
| Plastics: 250 lb | 300 kWh | |
| Paint: 4,700 sq ft | 500 kWh | |
| Copper and brass: 500 lb | 2,500 kWh | |
| Others: | 4,000 kWh | |
| | 31,510 kWh | |

*Materials Transport*

| | |
|---|---|
| Bricks 60 miles at 1·5 kWh/ton mile: | 3,200 kWh |
| Timber 250 miles at 1 kWh/ton mile: | 1,100 kWh |
| Cement 40 miles at 1·5 kWh/ton mile: | 400 kWh |
| | 4,700 kWh |

Total inputs

31,500 + 10,200 + 4,700 = 46,400 kWh

Alternative 1 : 10% soil cement blocks

| | |
|---|---|
| 8 cu yd cement and handling: | 12,500 kWh |
| Soil: 50 tons (hand labour): | 50 kWh |
| Localised wood supply: | 150 kWh |
| Glass: | 2,000 kWh |
| In situ rendering materials: | 100 kWh |
| Metals: | 1,500 kWh |
| Others: | 2,500 kWh |
| Total inputs: | 18,800 kWh |

Alternative 2: Rammed earth

| | |
|---|---|
| 80 cu yd earth. 70 men days: | 100 kWh |
| 160 cu yd earth invert: | 150 kWh |
| Glass: | 1,500 kWh |
| Timber: | 150 kWh |
| Rendering: | 50 kWh |
| Metals: | 1,000 kWh |
| Others: | 2,000 kWh |
| Total: | 4,950 kWh |

annual rental of £10 per square foot, shortly to increase to about £20 per square foot. The running costs of a moderately designed air-conditioning installation are 40–60p/sq ft and the depreciation costs are of the same order. Thus the financial incentive to reduce both capital and running costs and to conserve energy is virtually non-existent, as the marginal savings would amount to about 5%–10% of the rental.

There is a case for a swingeing penalty to be imposed on the designers of installations that do not use energy and materials in an optimum manner. The examples given previously of inefficient thermal insulation and excess power consumption in air-conditioning show that there is scope for great improvement in many areas. A surcharge should be levied on new installations that lead to excess energy consumption, so that the nation's interests are protected.

In an energy shortage the most efficient energy utilization systems should be adopted. We must also ask whether the present pattern of consumption is proper in providing a quality of life which may inflict hardship on future generations. Energy accountancy can help to evaluate the efficiency of a system and the quantity of energy consumed, but society must judge between short-term and long-term advantage. Two areas in which substantial energy savings might be made are: (*a*) the insulation of domestic and industrial premises; and (*b*) recycling.

*Domestic and industrial heating*

Conservation measures are generally judged by how much money can be saved—not by how much of the earth's precious resources can be left in the ground. It is not generally realized that fully 40% of total UK energy

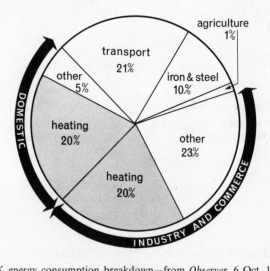

**Figure 3.5** UK energy consumption breakdown—from *Observer*, 6 Oct. 1974 (compiled from energy reports listed in references).

consumption is accounted for by domestic heating, as shown in figure 3.5. Many reports have drawn attention to the saving that can be effected by better home insulation.[9,10] The consensus is that a reduction in energy consumption of at least 40% can be effected in 8–11 million of Britain's 18,000,000 households if a programme of full thermal roof insulation (50 mm thickness), were realized.

The arithmetic is done as follows. The Institution of Heating and

Ventilating Engineers[11] estimates that 50 mm of mineral wool in the roof of a 3-bedroom semi-detached house at a cost of £37 will save £40 per year. On 11,000,000 houses this is a saving of £440,000,000 per annum. Cavity-wall insulation at a cost of £75 a house can save about £20 a year on the 8,000,000 houses suitable for such treatment, i.e. another £160,000,000 each year. Draught exclusion is difficult to quantify, but the Rothschild report estimated up to 30% heat loss due to draughts in some houses, say £15 saving per house on 18,000,000 houses, i.e. £270,000,000. Annual savings on domestic fuel bills would be on present rates around £870,000,000. The massive capital investment required would need to be spread over several years or decades, but the incentive is clearly demonstrated.

In industry the problems are less evident. The energy cost per unit of product may be only a fraction of the product's total cost; and in relation to other items required, e.g. machine tools and raw materials, the energy cost is still not a very substantial component. Energy conservation to many business men still has a low priority, and any increase in energy cost is passed on to the consumer. Yet one author[12] has estimated that the combination of poorly maintained steam lines and start-up practices gives rise to 27% excess steam consumption. Factory heating controls also need improving. Where heating is the sole use of fuel, 10% savings can usually be made, and the cost of automatic controls quickly recovered. One survey has shown that 4% of all industrial heating control systems were either inoperative or seriously defective.

Clearly a case can be made for energy saving, allied to improved domestic and industrial heating practices. Another area where a case can be made is that of recycling.

## Recycling

The natural recycling of the substances required for life is the keystone of our existence on earth. In the carbon cycle, the reuse of material is obtained with the aid of solar energy. We cannot bypass the conservation laws of mass or energy, and in a world of increasing scarcity it is important to make the best use of all resources. The intelligent adoption of recycling techniques allied with good design practice, which allows for materials to be reclaimed after their useful life is over, can do much to conserve raw materials and energy, minimize pollution, and save money. It would be too simplistic to expect people voluntarily to cut back their standard of living, so that major energy savings can be made, but recycling does offer the potential for considerable savings without major sacrifices on the part of the consumer. Recycling falls into three classes.

## (1) Reuse

Reuse is typified by the returnable bottle; it makes several trips from bottler to consumer and back again, where it is cleaned and refilled. Reuse may be allocated the highest availability in the recycling spectrum, in that least energy and process complexity is normally expended in getting the material (article) back into use.

## (2) Direct recycling

Using the returnable bottle as our link example, once it is unfit for reuse it may be cleaned and broken down to cullet at the glassworks and used to make more bottles. Direct recycling is dependent on the quality of the recycled material and on its cost, which should not exceed that of the virgin raw material. Currently most direct recycling occurs where the product is made, e.g. misshaped or broken bottles formed during glass manufacture are fed back to the melting chamber. This form of recycling is not to be confused with reclamation of material from waste or at the point of use. Thus paper with a 20% recycled content may in fact be paper where surplus pulp fibres, mill offcuts and spoiled rolls have been internally rerouted back through the pulping process. Direct recycling has an intermediate availability in that both energy expenditure and process complexity may be required in getting the material back into use.

## (3) Indirect recycling

Indirect recycling often makes no pretence at reclaiming the material for use as such, but rather gets a second bite at the cherry. Continuing with our glass bottle, it is quite probable that it will eventually end up in domestic refuse, where it can be extracted by screening and separation in conjunction with other bottles. These bottles will probably be of different colours and varying degrees of cleanliness, and are unsuitable for cullet use unless costly optical sorting is used. (The reclamation of glass for remelting from refuse has recently been shown to be more expensive in Britain—because of transport and collection costs—than the extraction of the basic raw materials. In the United States, optical grading of glass is under active consideration.) The bottles may, however, be ground up and used for a highly skid-resistant and durable road surfacing material. Waste plastic containers which are en masse unsuitable for direct conversion to new containers, may be ground up and used for plastic fence posts, pallets and chipboards, where appearance and structure are not primary considerations. Other examples of indirect recycling are (a) the conversion of refuse to combustible gases, (b) the heat of combustion used for district heating by means of incineration with heat recovery.

This form has the lowest availability in the recycling spectrum; once processed in this phase the material is no longer available for use except for landfill or incineration. The downgrading in use of several typical products is shown in Table 3.5.

Table 3.5 Downgrading in indirect recycling

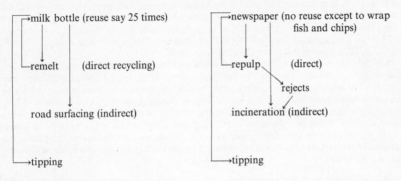

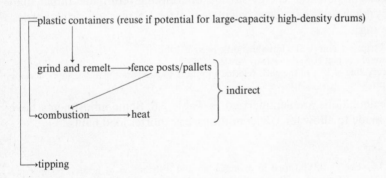

Tipping is not a form of recycling; it is a sink for discarded materials, just as our surroundings form the last resting place for degraded energy.

*Returnables v. disposables*
The glass bottle industry in the United Kingdom accounts for 0·4% of energy consumption—not in itself a large proportion—but this still contributes to the total energy consumption, and any energy savings should not be dismissed. There is a case for abandoning the one-trip philosophy and by implication the design ethos which prescribes the discarding of washing machines or cars after 4–6 years of use.

The complexities and market forces that militate against the adoption of returnable glass containers are highlighted in the statement below[13] by Mr R. F. Cook, Marketing Manager of the Glass Container Division, United Glass, commenting on the potential threat of one-trip plastic milk containers:

The distribution of milk is not a free-market situation. The government sets policies and controls prices to maintain a high level of liquid milk consumption, most of which is delivered daily to the door. Present arrangements for delivery make the use of returnable glass bottles both practical and economic, and 84% of all liquid milk is sold in this type of container. In 1971 the glass industry supplied 455 million bottles to dairies, each one costing approximately 1·75p, a price that has not changed since 1962.

However, economic use of glass bottles relies on their return for reuse, and the trippage is an important economic factor. The national average for return of bottles is 25 trips, but there is a variation across the country with the highest return rate (average 40 trips) along the south coast, compared with the lowest rate of 5 trips in the North of Scotland. More disturbingly, there is a gradual decrease in the number of trips back to the dairy over the whole country. When the number of trips falls below about seven, it begins to be as cheap to use non-returnable containers, and as soon as large-scale non-returnable packaging appears, supermarkets and other retail outlets, who are loath to handle returnable containers, become interested.

While this was a comment on one-way plastic containers, the extreme reluctance of supermarkets to adopt returnable packaging is also highlighted. The potential energy savings of adopting returnable rather than disposable bottles have been calculated[14] for US pint soft-drink bottles, assuming:

(1) 8 trips per pint (US) returnable bottle (weight 1 lb);
(2) weight of pint (US) throwaway bottle = 0·65 lb;
(3) recycling of 30% of the glass rejected to refuse. by a separation and colour-checking process.

The calculations are summarized in Table 3.6. Some adjustments have been made to allow for UK conditions, e.g. shorter load hauls.

**Table 3.6** Energy ($kWh_t$) used in manufacture and distribution of bottles: comparison between 8 × 20-oz throwaway and 1 × 20-oz used eight times

|  | Returnable (8 fills) | Throwaway |
|---|---|---|
| Raw material winning | 0·36 | 1·90 |
| Transportation of raw materials (½ US distance) | 0·02 | 0·09 |
| Manufacture of container | 2·83 | 14·93 |
| Manufacture of cap | 0·57 | 0·57 |
| Transportation to bottler (½ US) | 0·05 | 0·26 |
| Bottling (US) (includes rinsing) | 1·79 | 1·79 |
| Transportation to retailer (⅓ US figs.) | 0·17 | 0·12 |
|  | 5·79 | 19·66 |

5·79 $kWh_t$ is consumed for 1 US gal of drinks in returnable bottles and 19·7 $kWh_t$ is consumed for 1 US gal of drinks in throwaway bottles—a ratio of 3·4. When the trippage is increased to 16, and similar calculations performed, the energy ratio becomes 4·76.

Now the UK picture is such that disposable glass containers outnumber the returnables by about 4:1, and it could be asserted that the energy costs incurred by picking up loads of empty returnable bottles and returning them to their point of use outweigh the energy costs of using disposable bottles. This may be the case, but we may be pardoned for suspecting that the reluctance of supermarkets to use returnable bottles is a major factor in their not being adopted on a larger scale. (One well-known cider manufacturer has returnable bottles at off-licences and the same-capacity non-returnables for supermarkets.) Hannon's calculations show how energy accountancy can be put to use, and how areas of potential-energy savings are highlighted. In fact we would go further and suggest that glass containers could be standardized in two classes:

(1) wide-mouthed screw tops for pastes, etc.;
(2) narrow-mouthed for liquids.

The capacities could be $\frac{1}{4}$ litre, $\frac{1}{2}$ l, 1 l, and the shape and colour uniform throughout the industry. Thus drink manufacturer A may sometimes use manufacturer B's bottles—but it is their contents that people want, not the bottles.

Table 3.7 gives the energy consumption in tons of oil equivalent to produce 1 million containers (capacity $\frac{1}{3}$ litre) in various materials. It is obvious that the multi-trip returnable glass bottle has considerable energy savings over all other materials, and reinforces the contention of many conservationists that improvement means deterioration. Materials substitution can be wasteful of both energy and raw materials.

**Table 3.7**  Energy to produce 1 million $\frac{1}{3}$-litre containers[15]

|  | Tons of oil equivalent |
|---|---|
| Returnable glass (1 trip) | 138 |
| All-aluminium can | 117 |
| 35-g plastics bottle | 110 |
| Non-returnable glass | 96 |
| Tinplate can (aluminium end) | 82 |
| 25-g plastics bottle | 81 |
| All-tinplate can | 64 |
| Returnable glass (8 trips) | 34 |
| Returnable glass (19 trips) | 25 |

*Materials substitution*

It is not uncommon for plastic containers to replace glass bottles (not usually for carbonated beverages). Now the total energy input in plastics manufacture (polyethylene) is about $16.5 \, kWh_t/kg$ and if a lightweight 20-oz plastic bottle weighs 10 g, then to replace a gallon (160 oz) of drink in returnable glass bottles (16 trips) requires 1·28 kg which requires an energy input of 21 $kWh_t$ compared with 5·49 $kWh_t$ for returnables.

So plastics substitution for glass is of doubtful advantage, and the disposal of the plastic containers has not been considered; the mind boggles at the thought of plastics replacing all the returnable milk bottles in existence—not only from an ecological and pollution viewpoint but also at the needless consumption of excess energy and oil in the interests of marketing.

## Domestic refuse

The many possibilities for utilizing waste materials are readily illustrated by considering domestic-refuse disposal in detail. To the local authority, domestic refuse is material which must be disposed of; in the United Kingdom it amounts to 18 million tons per year. The commonly used methods of disposal are tipping (80%) and incineration (20%). However, tip space is running out, and the adoption of incineration is increasing, to ensure volume reductions of roughly 90% as the space-consuming paper and plastics are totally obliterated in the process. However tipping and incineration do not improve the environment. Proposals—and in some cases actual processes—now exist to recycle constituents in refuse, reclaim energy or recover by-products. In the discussion that follows, the composition of refuse and its recovery potential will be analysed and the existing state of the art costed in comparison with incineration. Various recovery/recycling processes will be described and costed to a standard format to illustrate their viability against incineration as a competing alternative choice.

## Refuse composition and recovery potential

To understand the potential of refuse as a source of raw materials, it is usual to divide it into two chemical classifications: organic and inorganic. The organic materials are paper, rags, plastics and foodstuffs. The inorganic are glass, ashes and metals. The average refuse analysis by weight for the United Kingdom is given in Table 3.8.

**Table 3.8**  British rubbish

|  | Percentage (by weight) |
|---|---|
| Paper and cardboard | 38 |
| Vegetable and foodstuffs | 20 |
| Plastics | 1 |
| Rags | 2 |
| Dust and ashes | 17 |
| Glass | 10 |
| Metals | 10 |
| Unclassified | 2 |

This means that refuse consists on average of approximately 60% organic and 40% inorganic material.

Taking the inorganic materials first, metal can be recycled to provide a saleable ferrous scrap. Glass presents problems, as some may be coloured, and glass cullet must be graded by composition, cleanliness and colour. So the prospects for recycling glass from refuse on a large scale are poor. A pilot scheme in York, which relied on the willingness of the housewife to wash and clean all bottles and separate them into coloured and uncoloured, did not match the costs of extracting fresh raw materials (which for glass are in abundant supply). Ashes and dust are usually valueless and must be dumped.

Organic components are interesting. First, straight recovery can be practised, as in salvaging paper. Roughly $6\frac{1}{2}$ million tons of waste paper, packaging and board are thrown away in refuse, yet less than $\frac{1}{2}$ million tons are collected from domestic sources for repulping. Another $1\frac{1}{2}$ million tons are salvaged from trade premises, printers and paper mills. Why is the amount of paper recovered from domestic refuse so low? The President of the British Waste Paper Association said in 1967:

> The waste paper industry recognizes the important contribution that local authorities make towards their efforts, but at the same time it would like local authorities to recognize that over the years processors have developed sources for the disposal of their graded goods on quality standards which it would be impossible for local authorities to emulate—unless they were prepared to educate, train and equip special personnel. . . . The most common grades collected [from households] are old newspapers, magazines, boxboard and containers. These wastes generally have no commercial use in paper making, and a limited use in board making.

These remarks are still true today, with the proviso that board mills may pay more for suitable salvaged paper and board, as there was a projected demand for an additional input of around $\frac{1}{4}$ million tons annually to satisfy the current shortfall of virgin materials. It is to be hoped that this direct form of recycling will grow, but there is a limit to the amount of these raw materials that the mills can take, and substantial quantities of paper will still remain in the refuse. Not all the paper and board in refuse can, by any stretch of the imagination, end up as fodder for the board mills, but recycling up to this limit should be practised and actively encouraged. The complexities of the waste paper market are best demonstrated by the fact that there are 38 different grades of waste paper and board, and it is usual for paper from domestic refuse to be graded into only three of these categories.

What then can be done with the organic components in refuse? The first and most obvious one is heat recovery from incineration. Why send all that good heat up the chimney? Heat recovery or recuperation is accomplished by adding a boiler to an incinerator to generate steam, which can then be put to a variety of uses, including power generation, as in Amsterdam; desalination of sea water, as in Hong Kong; or district heating, as in Paris and Nottingham.

However, power generation has also been attempted in the United Kingdom. The giant Edmonton incinerator was built by the Greater London Council to incinerate 1300 tons of refuse a day and generate 25–35 megawatts of electricity for sale to the Central Electricity Generating Board. But the cardinal premise of any extra expenditure over and above straight incineration with no heat recovery, is that the revenue from the extra capital employed will ensure a net reduction in the refuse disposal costs. So, if ordinary incineration costs £x per ton, and incineration and power generation cost £(x + y) per ton, then the local authority would choose the first alternative. Unfortunately Edmonton has not performed according to plan and, as a result, the cause of power generation from incineration has received a severe blow.

But the basic concept of heat recovery as a form of recycling is still valid. As fuel costs increase, we may expect the adoption of recuperative schemes for district heating in Britain. With this method blocks of flats may be heated by hot water provided by steam produced at an incineration plant. The problem is that refuse composition varies from hour to hour, which means that heat recovery can fluctuate. Also, peak demand for district heating occurs in winter, and the refuse output does not increase to match it. Supplementary fuel must therefore be used. In summer, the heating load is decreased, but the incinerator must still burn the refuse. If there is a plant breakdown, blocks of flats may be without heating unless standby plant is available.

But district heating does work; Nottingham combines it with refuse disposal, and claims a cost reduction compared with conventional incineration. Nottingham has an industrial partner in the National Coal Board which sells the heat. Information about how profitable it is to both parties must be available before district heating can be correctly assessed. Because of the lack of cost information on district heating, incineration only will be analysed here to establish a cost norm that other processes must better if they are to be considered.

### Incineration

Incineration is the term used for the combustion of municipal refuse. In a properly designed and operated incinerator there is a substantial reduction in the volume of waste material to be disposed of by tipping.

The process is extremely hygienic, and many of the problems associated with controlled tipping (such as windblown refuse, rodents and flies) are eliminated. Properly incinerated refuse becomes a sterile ash, with minimal

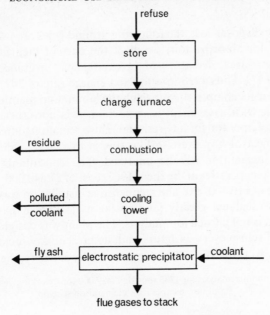

**Figure 3.6** Incineration flow diagram.

carbon or fat content, and thus can be safely tipped in almost any location. The greatest point in its favour is the substantial volume reduction which is about 90% of the original refuse volume; 60% weight reduction is not uncommon.

The basic elements of a non-recuperative incineration scheme are given in figure 3.6. It is seen that there can be four reject streams on which environmental control must be practised, namely the incinerator residue, fly ash, cooling water and the flue-gas stream. The flue-gas stream requires strict and costly control measures.

The predicted combustion performance of the Edmonton (GLC) incinerator is given in Table 3.9.

**Table 3.9** Predicted combustion performance of Edmonton incinerator (Greater London Council)

| | |
|---|---|
| Crude refuse input | 1333 tons per day |
| Output: clinker | 259 tons per day |
| metals | 147 tons per day |
| fly ash | 75 tons per day |
| Total residuals | 481 tons per day |
| Percentage weight reduction | 64 |

## Cost assessment

To assess the disposal cost, the following ground rules are used. A rate of 17% is used for amortization to reflect the current high interest rates, municipal rates are levied at 2% of the capital cost, insurance and inspection services 1%. Thus a composite fixed charge rate of 20% is obtained. The fixed charges completely outweigh the labour and maintenance costs, so that, while their assessment is important, it is not crucial. The plant is also levied £2 per ton for any residual refuse remaining after processing.

As the object of any exercise like this is to reflect reality, the GLC Edmonton incinerator will be analysed. The consultants costed two versions: (a) recuperative at the then (1967) cost of £7,640,000; and (b) non-recuperative of £10,400,000. The recuperative installation was chosen but, as the costs escalated greatly (and so far as is known have not been published), it is not chosen for analysis. The plant was designed to handle 1333 tons of refuse every 24 hours and, as present-day incinerator costs approximate to £10,000–£12,000 per ton of daily capacity, the lower figure

**Table 3.10** Economic analysis for 1333 tons/day (487,000 tons/working year) crude refuse input non-recuperative incineration scheme

| | |
|---|---:|
| Plant cost | £13,330,000 |
| *Annual cost* | |
| Fixed charges at 20% | 2,660,000 |
| Wages (estimated) | 250,000 |
| Electricity | 200,000 |
| Disposal charge for residual material at £2/ton and mean net rate of 334 tons/day | 244,000 |
| Total gross annual cost | £3,354,000 |
| Less revenue from scrap sales | 100,000 |
| Net annual cost | £3,254,000 |

$$\text{Disposal cost per ton of input refuse} = \frac{3 \cdot 254}{0 \cdot 487} = £6 \cdot 68$$

will be used. The published figures (1967) for labour rates are updated. Table 3.10 summarizes the results of the economic analysis.

Thus the costs incurred in the adoption of incineration are not negligible, and at a conservative £6·68 per ton provide a cost norm which the alternative processes must equal or better.

## Refuse pyrolysis for gas production

In the processing of refuse or, for that matter, in the solution of any waste-disposal problem, there must be no conversion of pollution from one form into another. Incineration of waste can give rise to air pollution, and hydrolysis of refuse can give rise to water pollution, but both forms of pollution can be tackled provided strict control is maintained. One process which appears to have a strong potential for producing both usable end products and effecting a readily controlled disposal is *pyrolysis*.

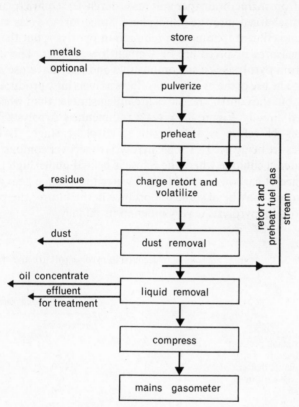

**Figure 3.7** Pyrolysis flow diagram.

Pyrolysis or destructive distillation of refuse is the indirect heating of the refuse in a retort (i.e. an enclosure capable of withstanding high temperatures) at elevated temperatures in the range 250–1000°C, depending on the processing techniques employed and the products desired. Virtually any carbonaceous material that can be volatilized can be pyrolized. From the refuse disposal viewpoint, the wastes can be converted to usable or more manageable solid, liquid and gaseous forms with less potential for contributing to land, air and water pollution (provided the effluent receives proper treatment).

The main reason for this enhanced environmental acceptability is that pyrolysis is not a combustion process and is carried out in closed vessels or retorts; thus the much larger air volumes, fumes, grit and dust loadings associated with incineration are not present. The gases from pyrolysis are combustible and can be used as fuel, whereas the much greater volume of flue gases from incineration (per unit mass of refuse) require treatment and are then discharged into the atmosphere. Some form of gas cleaning or washing and effluent treatment is required in pyrolysis, but the pollution control measures involved are of a much lower order. The char (solid residue) from pyrolysis is also combustible and may have use as a fuel if upgraded. The use of the pyrolysis products as fuels may produce pollution problems, but the control measures are much simpler than when refuse is incinerated directly. Figure 3.7 gives a rudimentary pyrolysis flow sheet.

Pyrolysis of refuse is still at the pilot-plant stage. Two distinct approaches are being used—simple pyrolysis to recover combustible gases, and a related technique where the refuse is heated under high pressure in the presence of carbon monoxide and water to produce heavy oils. The former process may be of some importance in the future; the production of heavy oils by pyrolysis is very much in its infancy.

Table 3.11  Average analyses of refuse used in pyrolysis (destructive distillation) tests[16]

|  | Raw municipal refuse | |
|---|---|---|
|  | as received | dry |
| Proximate, per cent: | | |
| Moisture | 43·3 | — |
| Volatile matter | 43·0 | 76·3 |
| Fixed carbon | 6·7 | 11·7 |
| Ash | 7·0 | 12·0 |
| Total | 100·0 | 100·0 |
| Ultimate, per cent: | | |
| Hydrogen | 8·2 | 6·0 |
| Carbon | 27·2 | 47·6 |
| Nitrogen | 0·7 | 1·2 |
| Oxygen | 56·8 | 32·9 |
| Sulphur | 0·1 | 0·3 |
| Ash | 7·0 | 12·0 |
| Total | 100·0 | 100·0 |
| Btu per pound of refuse | 4,827 | 8,546 |
| kJ/kg of refuse | 11,200 | 19,800 |
| Available Btu per ton of refuse, millions | 9·654 | 17·092 |
| Available kJ/kg refuse | 9,900 | 17,500 |

*Pyrolysis for gas production*

The refuse is heated in a retort, yielding solid, gaseous and liquid fractions. Substantial work is being conducted in the United States; the analysis of refuse used in some recent tests is given in Table 3.11.

Pyrolysis was carried out at two constant temperatures of 750°C and 900°C. The products and their yields are given in Table 3.12.

**Table 3.12**  Yield of products from pyrolysis for gas of 1 ton of treated US refuse

| | | |
|---|---|---|
| Char | 154–230 lb | 70–104 kg |
| Tar and pitch | 0·5–5 US gal | 1·9–19 litres |
| Light oil | 1·5–2 US gal | 5·65–7·5 litres |
| Ammonium sulphate | 18–25 lb | 8·15–11·3 kg |
| Liquor | 80–133 US gal | 302–533 litres |
| Gas (15°C, $1 \times 10^5$ Pa) | 11,000–17,000 cu ft | 314–486 m³ |

The char obtained had characteristics which varied widely depending on the refuse. Municipal refuse with a high plastics content yielded a char with 6·7% fixed carbon, whereas industrial refuse produced a char with roughly 12% fixed carbon. The sulphur content was less than 0·2% in all tests.

The gas content is most interesting. The major constituents are hydrogen, methane, carbon dioxide, carbon monoxide and ethylene, in that order. An analysis is given in Table 3.13.

**Table 3.13**  Gas analysis from pyrolysis of municipal refuse

| Pyrolysis temperature (°C) | 750 | 900 |
|---|---|---|
| | *percentage volumes* | |
| Hydrogen | 31 | 52 |
| Carbon monoxide | 16 | 18 |
| Methane | 23 | 13 |
| Ethylene | 11 | 5 |
| Carbon dioxide | 19 | 12 |
| Btu/cubic foot gas | 563 | 447 |
| MJ/m³ | 21 | 16·7 |
| Million Btu/ton refuse pyrolised | 5·421 | 7·930 |
| kJ/kg refuse pyrolised | 5,600 | 8,200 |

The gas analysis in Table 3.12 is in line with the pyrolysis work reported by the Karl Kroyer Company[17] in Denmark.

The yield of combustible gas is higher at 900°C than at 750°C. Now the energy requirement for the pyrolysis of 1 ton of refuse is roughly $2 \cdot 1 \times 10^6$ kJ (2 million Btu); and as approximately $8 \cdot 2 \times 10^6$ kJ (8 million Btu) are available per ton at 900°C the process is self-sustaining in

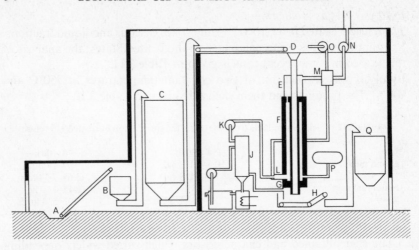

**Figure 3.8** Pyrolysis process as installed at Kolding (Destrugas process, Denmark).

energy requirements and has a surplus which can be used to augment town gas supplies; this has been done in the Danish town of Kolding in the 'Destrugas' pilot plant operation.[17] Figure 3·8 shows a schematic diagram of the Kolding plant.

Gas production by refuse pyrolysis is now available commercially, and it will be interesting to see the extent to which this method is adopted. An economic analysis based on the probable maximum yields from UK refuse is given in Table 3.14. It is to be noted that (a) the selling price of the gas has a marked effect on the return, and (b) the capital cost of the plant is assumed to be equal to that of a comparable incineration instal-

**Table 3.14** Economic analysis for pyrolysis plant producing (in gaseous form) $4 \times 10^6$ Btu/ton of refuse for sale at 50p/$10^6$ Btu

| | |
|---|---|
| Plant cost/ton daily capacity | £10,000 |
| *Annual costs (for input of 1 ton/day)* | |
| Fixed charges at 20% | 2,000 |
| Wages | 200 |
| Electricity | 200 |
| Disposal charge: 100 tons/year at £2/ton | 200 |
| Total cost | £2,600 |
| Less revenue from gas sales at 50p/$10^6$ Btu | 730 |
| | £1,870 |

Disposal cost = 1870/365 = £5·1 per ton

lation. Should the costs exceed this it is doubtful if pyrolysis will be attractive.

Note, however, that the gas calorific value is roughly half that for natural gas, therefore the market may be highly restrictive to captive customers who can demand a lower selling price.

*Hydrolysis of cellulose*

Cellulose is nature's principal polymer; it is the chief structural element and major constituent of trees and higher plants. It has the empirical formula $C_6H_{10}O_5$ and is available almost globally.

As solar energy is used to convert water and carbon dioxide into plant material by photosynthesis, a virtually unlimited source of raw material is available. The use of cellulose via the process of hydrolysis to produce protein and/or ethyl alcohol will assume great importance in the future as the income of solar energy can be utilized to provide food (energy) for a protein-short world.

There are many wastes that can be used to provide protein. Forster and Hughes[18] have estimated that the soluble sugars in the spent sulphite liquor from the processing of $3 \times 10^6$ tons of wood pulp produced annually in the United States has a protein potential of $1 \cdot 5 \times 10^5$ tons with a value of £45 $\times 10^6$. Wastes containing cellulose occur wherever crops or timber are grown and processed. Thus the utilization of this 'free' source of carbohydrate is of the utmost importance.

Gray and Biddlestone[19] have identified substantial quantities of organic wastes in the United Kingdom as summarized in Table 3.15. All these wastes are suitable for hydrolysis and can be converted by microbial action under the proper conditions to yield proteinaceous material for animal feed or other fermentation products.

**Table 3.15** Approximate production of organic waste in the United Kingdom

| Source | Tons per year, fresh weight | Moisture content %, fresh weight basis |
|---|---|---|
| Wood—shavings and sawdust | 1,070,000 | ? |
| Straws—wheat, barley, oats | 1,000,000 | 14 |
| Potato and pea haulms, sugar beet tops | 1,600,000 | 77–83 |
| Bracken, potentially available | 1,000,000 | ? |
| Seaweed, potentially available | 1,000,000 | ? |
| Garden wastes | 1–10,000,000 | ? |
| Sewage sludge | 20,000,000 | 95 |
| Municipal refuse | 18,000,000 | 20–40 |
| Farm manures | 120,000,000 | 85 |

The paper and vegetable content of domestic refuse alone is capable of yielding about five million tons of cellulose a year. Cellulose can be processed to yield ethyl alcohol—a major industrial commodity currently derived almost exclusively from non-renewable oil reserves.

Hydrolysis of cellulose can be accomplished with the aid of acid as the hydrolysis reagent as follows:

$$C_6H_{10}O_5 + H_2O \xrightarrow[H_2O]{H_2SO_4} C_6H_{12}O_6$$

cellulose                    sugars

The acid acts as a catalyst, and for fast industrial reaction and high yields the process is carried out at elevated temperature. Once the sugars are available, a variety of operations can be carried out using various strains of yeast to yield many fermentation products, including yeast, vitamins, protein, fat, alcohol, acetone, glycerol and organic acids. The reaction for the production of ethyl alcohol from the sugar solution is:

$$C_6H_{12}O_6 \xrightarrow{yeast} 2C_2H_5OH + 2CO_2$$

sugars                      ethanol

The fermentable sugars change, on continued exposure to the hot dilute acid, into non-fermentable sugars of little industrial value. Thus for any given hydrolysis conditions (i.e. acid concentration and temperature) there is an optimum reaction time for maximum fermentable sugar yield, after which the temperature must be sharply reduced to quench the reaction and stabilize the yield. It can be shown that for high sugar yields and low costs the reaction should be carried out at high temperature and low acid concentration.

Hydrolysis conditions of 0·4% $H_2SO_4$ concentration and 230°C were found to be the approximate upper limits for a controllable reaction. The optimum residence time in the chemical reactor for maximum yield under these conditions is 1·2 minutes with a 55% conversion to fermentable sugars.

*Process description*
The hydrolysis process is conceptually quite simple. Figure 3.9 gives a rudimentary flow diagram for the processing of municipal refuse and illustrates the principal steps employed. Briefly, the refuse is pulverized and discharged into a flotation separator or a special pulper to allow segregation of the refuse into a dense and light fraction. It should be noted that air classification is also possible, which has the advantage of eliminating a waste water stream. The pulped fraction, mainly cellulosic material, is then fed to a fines and plastic removal section, and thence to a reactor for hydrolysis at 230°C and 0·4% $H_2SO_4$ with an optimum residence time

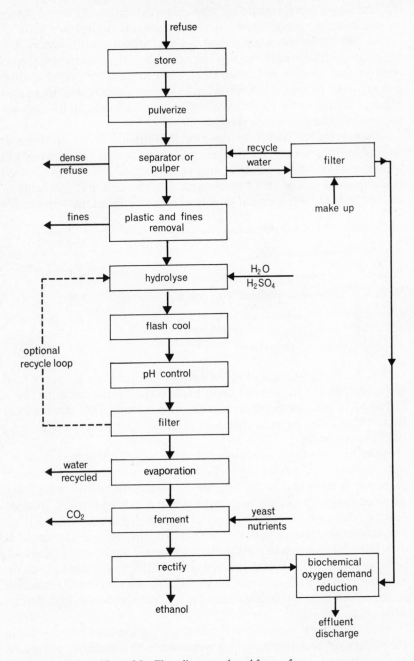

**Figure 3.9**   Flow diagram ethanol from refuse.

of 1·2 minutes to obtain maximum conversion to fermentable sugars. Flash cooling using the process feed water as coolant, pH control to a value of 5, and filtering follow. Evaporation is then carried out (to increase the sugar concentration) followed by fermentation. The resulting aqueous ethyl alcohol solution is distilled or rectified to yield 95% ethyl alcohol and 5% water by volume. A waste liquid stream which is also discharged requires treatment for biochemical oxygen-demand reduction—and is a potential source for protein production.

This process has been fully reported by Porteous. [20,21] In order to understand its economic and resource recovery potential, two analyses will be carried out for 40% and 60% paper content in the refuse. It is assumed that paper consists of 75% hydrolyzable cellulose and a 25% soluble or readily hydrolyzable portion which does not convert to fermentable

**Table 3.16**   100% ethanol yield calculations for a crude refuse input of 250 tons per day

| | | |
|---|---|---|
| Paper content, per cent, in refuse | 40 | 60 |
| Paper input tons/day | 100 | 150 |
| Cellulose (75% of paper) tons/day | 75 | 112 |
| Maximum available sugar = cellulose $\times \dfrac{180}{162}$ tons/day | 83·5 | 124 |
| Net sugar yield at 55% conversion, tons/day | 46 | 67 |
| 100% ethanol yield from sugars, evaluated from fermentation equation = sugar $\times \dfrac{92}{180}$ tons/day | 23·5 | 35·3 |

sugars. A 55% conversion of the cellulose to fermentable sugars is used in the analysis, and the fermentability of the sugars is assumed to be 100%.

Table 3.16 itemizes the stages in calculating the final 100% ethanol yield from 40% and 60% paper content for a plant handling a crude refuse input of 250 tons per day.

In order to assess the amount of solid refuse which will require disposal, mass balances are carried out in Table 3.16 on a dry basis. The volume reduction is estimated to be 70%. In computing Table 3.16 it has been assumed that a 20-ton/day non-paper group is carried through into the reactor. This group is easily hydrolyzable vegetable matter which will go into solution but is not assumed at present to contribute to the fermentable sugar content; pilot-plant work will show whether this is actually the case.

From Table 3.17 it is seen that the weight reduction is substantial; the volume reduction should be of order 70%, as the most bulky materials in the refuse have either been used up or discharged as a high-density filter cake.

As the hydrolysis of cellulose is now receiving considerable scrutiny in the world's research institutes, it is worth while to summarize the results of experimental work carried out by the Thayer School of Engineering in the United States.[22] The major conclusions from this stage of the work are:

1. The kinetics and yields of cellulose hydrolysis are very close to those originally predicted.
2. A large change in refuse composition has no appreciable effect either on the rate of fermentation or the yield of ethanol.

As the input refuse feedstock composition may change daily, these results refute a major criticism which has been levelled at the process. However, the potential of hydrolysis has been demonstrated quite clearly in the difficult area of domestic refuse, and its application to these feedstocks (such as agricultural waste, sawdust and wood chips) should pose few problems.

Table 3.17 Refuse mass balances (dry basis)

| | 40% paper | | 60% paper | |
| --- | --- | --- | --- | --- |
| | Solid tonne/day | Liquid tonne/day | Solid tonne/day | Liquid tonne/day |
| Input refuse | 250 | | 250 | |
| Separated out | 130 | | 80 | |
| Vegetable hydrolyzable group carried through into reactor | | 20 | | 20 |
| Readily hydrolyzed paper fraction | | 25 | | 37·5 |
| Unhydrolyzed cellulose (23% of gross cellulose—this may be recycled) | 18 | | 26 | |
| Cellulose converted to fermentable sugars (55%) | | 41 | | 62 |
| Cellulose converted to decomposed sugars | | 16 | | 24·5 |
| Totals | 148 | 102 | 106 | 144 |
| Total dry solids for disposal by landfill | 148 | | 106 | |
| Per cent weight reduction | 40·7 | | 57·5 | |

The plant economics are summarized in Table 3.18. The plant capital cost has been detailed and is comparable with incinerator installations. The materials have been costed liberally to allow for escalation. The plant is also charged for effluent treatment, although this may actually cover its costs by the installation of filters and spray driers to obtain animal feedstuffs—now a common distillery practice.

At 40% paper content in refuse, the refuse disposal cost is estimated at £1·07/ton which is to be compared with that of incineration at £6·68 per ton, i.e. a net credit of £5·61/ton when the cost datum is set by incineration. At 60% paper content the plant yields a net profit of £4/ton which is a considerable financial inducement compared with any other competitor process. An additional point is that the fermentable sugar cost is £54·5/ton (40% paper content) and £37·8/ton (60% paper content) if the plant were required to break even—this is to be compared with the free market price of sugar solution which is in excess of £100/ton (net sugar). If the plant can take a credit of £6·68/ton of refuse, as it is usurping incineration, the sugar costs drop to £18·3/ton (40%) and £12·9/ton (60%). The free market price for unrefined sugar is now £200/ton (1975).

**Table 3.18**   Ethanol plant economics

| Item | 40% paper content cost £ | 60% paper content cost £ |
|---|---|---|
| Plant cost | 2,500,000 | 2,500,000 |
| *Annual cost* | | |
| Fixed charges at 20% | 500,000 | 500,000 |
| Wages | 60,000 | 60,000 |
| Materials (acid cost doubled to allow costing latitude) | 175,000 | 179,000 |
| Biological oxygen demand reduction treatment of liquid waste | 71,500 | 108,700 |
| Residue refuse disposal charge (at £2/ton) | 108,000 | 76,000 |
| Total annual cost | 914,000 | 923,000 |
| Ethanol revenue at £100/ton* | 816,000 | 1,290,000 |
| Annual loss/profit | (loss) 98,000 | (profit) 367,000 |
| Cost/profit/ton | (cost) £1·07 | (profit) £4·00 |

* Based on the early 1974 selling price; ethanol and its derivatives will probably be considerably more expensive by the time this volume is published. The production of single-cell protein from the fermentable sugar substrate can readily be done.

**Conclusion**

Energy is the prime mover of our way of life, and its use is governed by the laws of thermodynamics—once used, it cannot be returned.

(1) The concept of large-scale power stations as the basis of our electricity supply industry needs rethinking. There may be room for smaller units and the implementation of total-energy practices. The competition between energy supply industries is not necessarily beneficial to the country as a whole.

(2) Thermal insulation and draught-proofing of domestic and industrial premises is essential to stabilize and eventually reduce energy consumption for heating.

(3) Wasteful uses of energy (because it is a small fraction of final product cost or property rental value) should be discouraged by fiscal measures

based on model energy consumption rates for well-designed processes or systems and any excess consumption charged at swingeing rates to aid energy conservation.

(4) A policy for recycling or reuse should be implemented and should not tolerate (for example) the multiplicity of throwaway bottles and cans in a variety of materials. Containers should conform to official standards based on (a) consumption of materials and energy in manufacture, (b) reuse potential, (c) potential for recycling the materials as construction, (d) potential for environmental insult if indiscriminately discarded or jettisoned to the refuse disposal system.

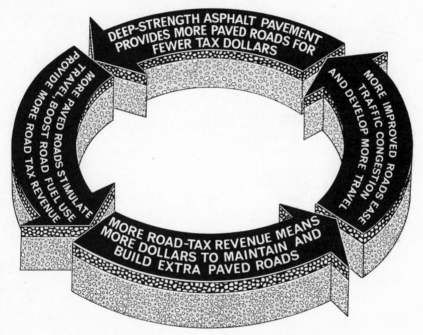

**Figure 3.10**   Asphalt's magic circle—source Asphalt Institute.

(5) Domestic refuse and other wastes are not something to be disposed of, but resources which can provide raw materials for industry, energy for district heating or gas for industrial use. Recycling conserves materials and energy, and can minimize pollution. Even if some solutions (such as pyrolysis) initially appear more expensive, conservation of resources for future generations must be considered to provide a credit.

(6) Society must start to develop a lifestyle that lives on income and does not rely on capital resources. Growth for growth's sake must stop; the magic circle of growth (figure 3.10) must be broken.

## FURTHER READING

1. King Hubbert, M. (1969), 'Energy Resources', Chap. 8, *Resources and Man* (c), N.A.S., W. H. Freeman.
2. Cloud, P. E. (1968), 'Realities of Mineral Distribution', *Texas Quarterly*, Vol. 11, p. 103.
3. Friske, W. R. (1971), 'Extended Industrial Revolution and Climate Change EOS', American Geophysical Union, Vol. 52, No. 7, p. 505.
4. The Open University, 'The Man-Made World', Technology Foundation Course.
5. *Environment*, Vol. 14, No. 5, June 1972, p. 14.
6. *Efficient Energy Utilization*, NATO, 1974.
7. The Institution of Heating and Ventilating Engineers, *Energy and Buildings—towards a new policy*, 1974.
8. McCillop, A. (1972), 'Low Energy House', *Ecologist*, Vol. 2(12).
   new policy, 1974.
9. *Energy Conservation*, A Study by the Central Policy Review Staff; Chairman, Lord Rothschild, HMSO 1974.
10. *Energy and the Environment*, Royal Society of Arts, 1974.
11. The Institution of Heating and Ventilating Engineers, *Insulation of Roof Spaces*, 1974.
12. David, T. (1974), 'Why Burn Money in Wasted Steam?' *Process Eng.*, p. 53.
13. The Open University, 'Introduction to Materials', Course TS251, Unit 9, p. 36.
14. Hannon, B. N. (1972), 'Bottles, Cans, Energy', *Environment*, Vol. 14, No. 2, p. 11.
15. Bunt, B. P. (1974), 'The Energy Cost of Packaging', paper read to the Institute of Food Science and Technology, 1 Aug.
16. Sanner, W. S., *et al.* (1970), 'Conversion of Municipal and Industrial Refuse into Useful Materials by Pyrolysis', USBM Solid Waste Programme Report, No. 7428.
17. Kroyer, Karl (1970), *Pyrolysis of Waste*, Stads-og havneingeniøren (Danish Publication).
18. Forster, C. E. and Hughes, D. E. (1974), 'Waste—a Potential Raw Material', Symp. Waste Recovery and Recycle, IChemE, London.
19. Gray, K. R. and Biddlestone, A. J. (1973), *Chem. Eng.*, No. 270–71.
20. Porteous, A. (1969), 'The Recovery of Industrial Ethanol from Paper in Waste', *Chemistry and Industry*.
21. Porteous, A. (1974), 'The Recovery of Ethyl Alcohol/Protein by Hydrolysis of Domestic Refuse', Inst. Solid Wastes Management, Symposium, Manchester.
22. Karandikar, S. M. (1971), 'Fermentations of Hydrolysed Solid Wastes', MSc Thesis, Thayer School of Engineering, Dartmouth College, Hanover, New Hampshire.

# CHAPTER FOUR

# ENERGY—THE INTERNATIONAL SCENE

SIR FRANK McFADZEAN

Although the rate of economic growth since the end of the second world war has varied widely from country to country, the free world as a whole has enjoyed, over more than two and a half decades, an increase in material prosperity unparalleled in its history. Many factors have contributed. The International Monetary Fund, designed to smooth transactions across the currency exchanges, and the World Bank, created to increase and ease the flow of capital from the richer to the poorer nations, were the outcome of the Bretton Woods negotiations towards the end of the second world war. Some time later, in Geneva, the General Agreement on Tariffs and Trade set the scene for a more open world economy. Together these agreements created the institutional framework for the subsequent advance in living standards; rapid technological progress played a major role; but the achievements would not have been possible without an adequate supply of relatively cheap energy. The supply-and-demand picture for primary commercial energy in the world, outside the Communist Areas, at various years between 1945 and 1973, is shown in Table 4.1. Coal, which had fired the industrial revolution and had been the dominant supplier of commercial energy up to the outbreak of the second world war, resumed that role in the early years of peace. As the demand for energy increased, the relative contribution of coal to the total began to decline as petroleum fuels and natural gas pre-empted the bulk of the rising demand. By the late sixties coal was declining, not only percentage-wise but also in absolute terms.

As is shown in Table 4.2, the level of commercial energy consumption per capita varies widely from country to country.

103

**Table 4.1**  Primary commercial energy demand/supply in the world outside the communist
area (1945–1973)

(Million barrels a day oil equivalent)*

|                  | 1945 | 1950 | 1955 | 1960 | 1965 | 1970 | 1973 |
|------------------|------|------|------|------|------|------|------|
| Oil              | 6·9  | 10·1 | 14·6 | 19·2 | 27·0 | 41·0 | 48·4 |
| Coal             | 13·0 | 14·9 | 15·4 | 14·6 | 15·8 | 15·5 | 15·7 |
| Natural gas      | 2·0  | 3·0  | 4·7  | 6·4  | 8·5  | 13·0 | 16·1 |
| Hydro-electricity| 1·8  | 2·5  | 2·9  | 3·7  | 4·5  | 5·0  | 5·8  |
| Nuclear          | —    | —    | —    | —    | 0·1  | 0·4  | 0·9  |
|                  |      |      |      |      |      |      |      |
| TOTAL            | 23·7 | 30·5 | 37·6 | 43·9 | 55·9 | 74·9 | 86·9 |

* The various forms of energy have, of necessity, been brought to a common base, and
the table is in millions of barrels per day oil equivalent. Conversion factors are:

1 million barrels per day of oil = 50 million metric tonnes oil per annum;
= 77 million metric tonnes of coal per annum;
= 5,800 million cubic feet of natural gas per day;
= $5·7 \times 10^{10}$ m$^3$ of natural gas per annum.

As would be expected, it is highest in the most advanced countries and
lowest in the so-called developing countries. However, the figures for the
latter tend to give a distorted view of the underlying position. Little or
no energy is required in tropical countries for the heating of homes or
factories. Cooking in many villages in Malaysia, for example, tends to be
by firewood collected by the villagers themselves; in agriculture, animals,
not tractors, tend to supply the energy and, in the classifications used, are
regarded as non-commercial. Again, as is shown in Table 4·3, the contributions of the various forms of primary energy also vary widely from
country to country.

**Table 4.2**  Per capita consumption of total primary
commercial energy in 1973

(Barrels of oil equivalent per capita)

| USA     | 63  |
|---------|-----|
| UK      | 30  |
| Japan   | 24  |
| Brazil  | 3   |
| India   | 1   |
| Nigeria | 0·5 |

This is due to the wide range of availabilities and to factors such as the
protection of indigenous supply sources for either security or social
reasons. Hydro-electricity plays a relatively large role in New Zealand
because of the availability of sites for water power; the finding of the large
Groningen gas field in Holland has resulted in gas assuming an important

Table 4.3   Percentage shares of total primary commercial energy in 1973

|  | Oil | Coal | Natural gas | Hydro-electricity | Nuclear |
|---|---|---|---|---|---|
| USA | 46 | 18 | 31 | 4 | 1 |
| Japan | 76 | 16 | 2 | 5 | 1 |
| UK | 51 | 35 | 11 | negligible | 3 |
| Netherlands | 54 | 4 | 41 | — | 1 |
| India | 31 | 56 | 1 | 11 | 1 |
| New Zealand | 37 | 10 | 3 | 50 | – |

part of the Dutch energy supply; while the protection of coal in the United Kingdom has played a large part in the primary energy profile of this country.

## The importance of oil

Since the age of steam, energy has had a global connotation. Coal moved within Europe and between continents on an appreciable scale, while the relative abilities of Germany and Great Britain to meet Italy's coal requirements was a major factor in the political and trade negotiations which preceded Italy's entry into the war in 1940. The post-war rise of oil as the pre-eminent contributor towards meeting the world's energy requirements increased the international overtones. Outside North America and the Soviet Union the principal producing oilfields of the world have been found mainly in areas of low consumption. The logistical requirements of having to move oil over vast distances either by tanker or pipeline has had considerable effect on the structure and investment patterns of the industry.

When the overall demand for refined products was relatively low, economics tended to dictate that a considerable part of world requirements could best be met by refining at the crude oil source; among other things this avoided hauling round the world the energy required to distil the crude oil into refined products. The growth in demand, the keen competition by host governments as well as by companies for crude oil outlets, the desire of many consuming countries for refineries within their own boundaries and, in certain instances, their preparedness to erect tariff barriers against products, tended to shift refining investment from producing to consuming areas. Moreover, the increasing size of ships reduced the penalty in transporting refinery fuel. The bulk of the international trade in oil from the fifties onwards took the form of crude oil movements with refined product shipments primarily concerned with meeting imbalances and seasonal variations in the different countries and continents.

Oil reached its preponderant position in the energy supply picture for several reasons. In certain end uses, such as the internal-combustion engine, there is little in the way of an alternative economic fuel to take its place; higher standards of living have resulted in more widespread car ownership. But even for uses, such as steam raising for heating and electricity generation, where it competes with other primary fuels such as coal, oil took a higher share of the market throughout the sixties. Oil affords a certain convenience of use compared with solid fuels, but the main ingredient in the changing market mix was its relative cheapness. Although obscured by the excise tax policies of various consumer governments, the oil industry during this decade not only contained inflation but reduced prices in the market place. This was due to many factors. The large and prolific fields in the Middle East enjoyed a very low technical cost of production and a relatively low investment requirement to expand output. There was therefore considerable pressure for the companies to produce the 'marginal' barrel. At the same time oil was discovered in other countries such as Nigeria and Libya. Although technical costs were appreciably higher than in the Middle East, some of the crudes enjoyed both a geographic advantage in respect of the rapidly growing European market and also a quality advantage in being low in sulphur content. When oil was found in countries which had not yet produced any, there was, quite understandably, pressure from host governments for early development and export to enable them to reap some of the financial benefits. Moreover, some of the companies which were successful, particularly in Libya, were relative newcomers to the international oil scene. Several of them were motivated by the desire to find a supplementary supply source for the peaking out and very high cost indigenous supplies in the United States. When the American Government imposed import controls to protect its local production, these companies set out to find a home for their 'orphan crude' in Europe. Competitive pressures were thereby considerably increased.

At the same time as these developments were taking place in crude oil production, the industry was effecting economies of scale in virtually all facets of the business. In the market place there was a growing concentration on high throughput stations, while unit costs were also reduced by a greater use of product pipelines and hydrant refuelling for aircraft. The size of distillation units in refineries was increased by factors of between three and five, while in marine transportation vessels of up to 50,000 tons were rapidly replaced by ships in the 200,000/350,000-ton range which effected savings of up to 70% in transportation costs. In addition to the use of larger units there was a rapid increase in the degree of automation; as a consequence, manning scales, particularly in refineries and ships, were reduced.

Although relative cheapness was the major consideration that resulted in oil exercising an increasingly important role in the energy market, there were other factors in the equation. The reserves proven by exploration world-wide, but particularly those in the Middle East, showed abundant supplies to meet the estimated requirements for many years ahead. This abundance of potential supplies, combined with the flexibility displayed by the industry in meeting successive crises, gave consuming countries a feeling of security, in the sense that they could rely on supplies not being interrupted to an extent that could cause grave economic damage. However, even before the events of 1973, the security of supply was becoming considerably weaker. In the Suez crisis of the fifties and sixties, the 'shut in' production in the United States, as a result of the pro-rationing policy, provided a reliable, if expensive, buffer to mitigate the adverse effects of a cut-back in other parts of the world. By the early seventies the expansion of oil demand in the United States had virtually eliminated the shut-in capacity. Again, the sheer magnitude of the volumes and the key position of certain producing countries such as Saudi Arabia and Iran, meant that a prolonged shut down of any one of them could produce substantial problems. Moreover, the accumulation by some producing countries of appreciable currency reserves meant that the economic pressure on these governments to resume production was much less than it was in the fifties.

The factors which benefited the consuming countries—the relatively low technical cost of production of crude oil, the economies of scale and the increased competition—evoked some considerable concern in the producing countries. The constant pressure on prices in the late fifties and early sixties resulted in the creation of the Organisation of Petroleum Exporting Countries (OPEC). Originally the prices 'posted' for crude oils, and on which the producing countries' 50/50 income tax was calculated, reflected the market realities, but from August 1960 this ceased to be so. OPEC action to prevent any downward movement effectively ossified the crude-oil posted-price structure for virtually a decade, and its only relevance was for the calculation of the host government 'take'. As prices in the market place fell, the tax arrangements remained nominally on the 50/50 basis but calculated, as they were, on an artificially high level, they became progressively more weighted in favour of the host governments due to their insulation from the sagging prices.

In an endeavour to raise prices OPEC in the middle sixties started to flirt with the idea of a pro-rationing scheme for oil production. Taking the twelve months ending June 1965 as a base period, the general idea was to forecast a rate of growth for five years ahead and allocate this growth among the various producing countries. All schemes of this nature embody a considerable degree of arbitrariness and most have within them the seeds of their own destruction. The OPEC proposal was no exception. The growth rate for the five years following the base period was put at 10%, while individual allocations ranged from a low of 3·3% for Venezuela—a well-established oil producer—to a high of 20% for Libya— a relative newcomer. The scheme was not a success, since the demand increase was not realized in practice and there was a considerable reluctance on the part of countries with new production to impose any restrictions on it if the concessionaires were able to find a market. But there was too little realization that a similar tactic could well succeed later with the growing imbalance between the industrial countries and the oil-producing states, the greater degree of sophistication being developed in the latter, and the easing of the pressures to produce as a result of increasing currency reserves.

### Economic and political factors

To understand the events of the early seventies it is necessary to recollect that for several Middle-Eastern countries crude oil is virtually the only substantial resource available. The technology of exploration and production, which developed it, was essentially Western in its origin. In an age of nationalism, the development of a natural resource by 'foreigners' built up political tensions which were probably aggravated in certain instances by the fact that a single company or single group of companies often had a concession covering either the whole or a major part of a country. This was largely the product of the particular era when the concessions were granted. At that time there were few enterprises with the resources available to risk exploration expenditure in remote and occasionally inhospitable areas. Moreover, the tools at the disposal of the prospector were primitive compared with modern techniques, and there was little in the way of knowledge about where in a vast area like Saudi Arabia oil was likely to be found. Whatever the reason, the single-company concession created a situation contrasting with the North Sea, where concessions were issued much later in a different economic and technological environment. As a consequence there are many companies involved in the North Sea, and individual blocks are relatively small. But in the Middle East with a single company developing the main, or only, resource, some governments tended to feel weak and, however unjustified it may have been, the fear of a state within the state was at least understandable. Although the alleged powers of multi-national enterprises *vis-a-vis* sovereign states belong largely to mythology, the oil companies had particular weaknesses in their negotiations with the governments of the producing countries. It would, of course, be incorrect to regard the OPEC members as a homogeneous group, but they did have more common purpose in achieving certain minimum aims than the industry which was fragmented in its approach. This was the result not only of American anti-trust laws but also the varying interests of the companies, particularly the 'independent' producers. Usually with only one substantial source of supply, they inevitably viewed government pressures in a different light from the majors with several sources. Whatever the independents were able to salvage from the main production area was a positive contribution to their overall position; they had no need to look over their shoulders at possible repercussions elsewhere.

In 1970 the OPEC Governments set out to increase their tax receipts. By that time Libya had accumulated sufficient currency reserves to make a substantial shut-in of production a matter of relative indifference. Motivated partly by a desire to secure what they regarded as a 'just price' for their oil, but also by the aim of showing the superiority of a radical government over the more traditional regimes of the Persian Gulf, Libya was among the most militant. By a skilful choice of tactics and a selection of the weaker companies first, Libya had succeeded by the end of 1970 in terminating the 50/50 profit-sharing agreements which had held sway in the Middle East for two decades and in jacking up the posted prices to levels even more remote from market realities than those previously prevailing. When the Persian Gulf producers effected comparable adjustments to their taxes and prices, the Libyans demanded further increases to restore the differentials. The prospect became one of the application of the ratchet principle with no clear views of where it would end.

The extra-territoriality of American anti-trust laws has long been an irritant to other governments. However, the consequences of a constant leapfrog in prices and taxes were sufficiently serious for a dispensation to be obtained from the United States Department of Justice to enable the oil companies to approach OPEC jointly. For various reasons, including internal political tensions, OPEC did not negotiate as a body. In February 1971 settlement was reached with the Persian Gulf producers in Teheran. This included an increase in the posted prices as well as changes in taxes, but more importantly, perhaps, a schedule of price increases was agreed up to 1975. Some three months later the principles of the Teheran agreement were extended to Libya and, with adjustments for geographic location and quality, to Nigeria and the Eastern Mediterranean. The agreements were freely entered into by the governments concerned, and it was hoped that they would provide a period of stability into the middle of the decade.

However, these hopes proved illusory, and uncertainty was again injected into the situation on two fronts. Firstly, the devaluation of the United States dollar, which is the currency in which most oil prices are expressed, resulted in a demand for 'ways and means to offset any adverse effects on the per barrel real income of Member Countries'. But secondly, and more importantly, a resolution was passed at the same September 1971 OPEC meeting to the effect that members should negotiate with the oil companies for 'effective participation in the existing oil concessions'. The latter resulted in lengthy negotiations which were completed with the signature of participation agreements in January 1973 between the international oil companies concerned and the major Arab oil-producing states in the Persian Gulf. Under the agreements the initial government stake was to be 25% and, although the companies failed to obtain compensation for oil in the ground, agreement was reached to pay 'up-dated book value'— a concept designed to make allowance for the increase in costs since the date of the initial investment. Participation was to continue at this level until January 1978 when governments could increase their participation to 30%, rising to 35% in 1979; 40% in 1980; 45% in 1981 and 51% in January 1982. Since the companies concerned had entered into complex supply arrangements, the governments agreed to help them 'bridge over' the initial difficulties by giving them access, at prices close to the market level, to 75% of the government's share in 1973, decreasing to 50% in 1974 and 25% in 1975. In addition to the 'bridging oil', there were complex provisions relating to 'phase-in oil'. This was oil derived from the government's equity interest and which they could 'put' on the companies if they were unable to sell it more advantageously elsewhere. In general the aim was to ease the government's oil on to the market in a non-disruptive fashion. In making the agreement to cover a substantial period of time it was once again hoped to introduce a degree of stability; once more the hope proved illusory.

As had happened with the price negotiations, Libya regarded the agreements reached in the Persian Gulf as establishing a floor only, and set about demanding stiffer terms as regards government participation, the level of buy-back prices and the compensation to be paid. The situation within Libya was confused. Towards the end of 1971 Libya nationalized BP's half-interest in the Sarir field, not because of any quarrel with BP but in retaliation for the alleged failure of the British Government to stop Iran occupying some small islands in the Persian Gulf. The other half-owner of the Sarir concession, an American called Bunker Hunt, was allowed to continue to produce at about 50% of the previous level but tension built up with the Libyan Government as Bunker Hunt refused to market the oil which had been nationalized. Since BP had not been paid the 'adequate, effective and prompt compensation' required on nationalization by international law, they pursued cargoes in the courts of some of the countries to which Libya sold the 'stolen oil'. Libya stepped up its pressure on Bunker Hunt, claiming half of his 50% of the Sarir concession, and also half the company's profits on its sale of crude oil since the time of BP's nationalization. In June 1973 Bunker Hunt was denied the right of export, and later that month, on the anniversary of the evacuation of the US Wheelus air base, Hunt's interests were nationalized to represent, as Colonel Gaddafi put it, 'a good hard slap' on the insolent face of America. Thus a British and an American company were nationalized, not for any shortcomings they possessed, but as a political gesture against their Governments—a process which was to be repeated yet again in 1974. Once more concentrating on the independent producers, who were responsible for about half of Libya's crude-oil exports, and under the threat that those who postponed reaching agreement would be given more onerous terms, the government forced participation up to a level of 51% immediately, while reducing compensation to net book value only—well below the 'adequate' level prescribed by international law—and forced the price of the government oil which the companies had to repurchase to a level well beyond what had been agreed in the Persian Gulf.

The possible repercussions of the Libyan events were crowded out by the outbreak of the Arab/Israeli war in October. The Arab Governments decided to use oil as a political weapon. Destination controls were imposed, production cuts were effected, accompanied by threats to deepen them by 5% in every subsequent month until 'the legal rights of the Palestinian people are restored', and government taxes—or government 'take' as it is more commonly known—were increased by something over 70%. Iran was a party to the price increase, but not to any threat of interruption of supplies. The agreements which had been designed to give some degree of stability on prices were unilaterally denounced, and any pretence at 'negotiations' with the companies was abandoned. The absence of any countervailing power on the part of the companies had progressively reduced them to a charade in any case. The OPEC Governments decided that they and they alone would in future determine the price at which crude oil was sold.

There can be little doubt that many of the OPEC Governments genuinely and sincerely thought that the main industrialized countries were receiving their oil too cheaply and that, in some way, the oil companies were responsible for this. It is incontrovertible that oil prices in the sixties were falling, contrary to movements in the general price level and the reasons for this have already been mentioned. It is equally incontrovertible that governments in most consuming countries had found oil, particularly gasoline, a convenient vehicle for extracting money from its people by way of excise duties. The oil-producing governments argued that the consuming governments were making more money out of their oil than they were themselves. In a sense this was true, but the deduction from it was erroneous. The Scots could argue the same way about their whisky exports to Japan and other countries. But it is the incidence of the tax that is the really important issue. The incidence of an excise tax, whether on gasoline or whisky, is on the citizens living within the jurisdiction of the government imposing it; the incidence of an export duty, particularly in the short run if no substitute is readily available, is on the citizens of the importing country. However, the heavy excise duties in most consuming countries on gasoline did show that demand was relatively inelastic.

The suspicion that the oil companies had sold oil too cheaply to the consuming countries was reinforced by the experience of some of the state corporations endeavouring to dispose of their participation crude. If they had entered the market in the sixties, they would have found, as the oil companies had, that it was definitely a buyers' one; all sorts of inducements had to be offered to secure outlets—long-credit terms, capital advances to independent refiners in exchange for a long-term crude-oil contract, and very keen prices. But they started marketing at a time when supplies were tight—the world-wide boom of 1973 had increased demand above anticipated levels, while the signs of imminent trouble in the Middle East had resulted in 'stocking up'. The situation where independent buyers could shop around for cheap marginal crude was reversed. The cut in Arab production and the threat of more to come resulted in panic buying; and the marginal price inevitably rose well above the price at which the bulk of the oil was moving through integrated channels. The marginal quantities were relatively small, but the OPEC Governments decided that the prices for these represented the real value of their crude oil. The situation was aggravated by the behaviour of some of the main consuming governments in the face of the crisis.

### Role of the multi-national corporation

In *The Age of Discontinuity* (Heinemann, 1969), Professor Drucker* observes: 'there is an . . . institution that the world economy needs; a producing and distributing institution that is not purely national in its economic operations and point of view. The world economy needs

* Clerk Professor of Social Sciences, Claremont Graduate School, California.

someone who represents its interests against all the partial and particular interests of its various members. . . . Traditionally such an institution has always been political, that is, a government. Because the world economy is strictly an economic community, the institution that represents it will have to be an economic rather than a political institution. Indeed it cannot possibly function unless it respects the political institutions of the nation state. The individual sovereign states, especially the big strong developed countries, will not accept a super government of any kind. Such an institution . . . we already have at hand. Its development during the last twenty years may well be the most significant event in the world economy and the one that, in the long run, will bring the greatest benefits. The institution is the "multinational" corporation.'

Professor Drucker wrote these words in 1968; by the autumn of 1973 the need for an institution to look after global, as opposed to purely national, interests became distressingly clear. In the face of the Arab cutbacks it was evident that the total world demand for oil could not be met, and the problem arose as to how the shortage should be distributed. It was of vital importance to the free world, since there was a threat that the screw would be continually tightened until the Arabs obtained their political objectives. It was a situation which called for international political leadership. The United States was almost precluded from the role by virtue of its Israeli policy. It was an opportunity for the political leaders of Britain and France; but it was one they failed to grasp. They failed to realize the essential interdependence of the world economy and the disruption that could follow the pursuit of chauvinistic policies. As the situation developed, there was both tragedy and comedy. The tragedy lay in the spectacle of convinced Europeans using various tactics to keep their own countries whole at the expense of deeper cuts in the supplies to others. It was not only tragic; it was rather naive. Indeed it bordered on the absurd to think that the Singapore Government, for example, if it suffered a disproportionately heavy reduction in supply would nevertheless be prepared to bunker British ships at the expense of fuel oil supplies to its power stations, while Britain's own supplies were kept intact. There were only a few countries that did not try by one means or another to unload at least part of the problem of world shortage on to the shoulders of others. Ships in the iron ore trade, which had traditionally taken on round-trip bunkers at their home ports, would arrive in Australia with one-way bunkers only, relying on the Government to pressurize the oil companies to supply bunkers for the return voyage to keep the ore moving.

The comedy was provided by the succession of people claiming to be able to obtain crude-oil supplies—at a price—by virtue of their direct or indirect access to this or that member of the government of an oil-producing state; and with the spectacle of two senior Cabinet Ministers flying off with the melodrama normally associated with the relief of a beleaguered fortress, to sign a barter agreement involving a quantity of crude oil equal to less than four weeks of the country's requirements. Delegations and emissaries, politicians and friends of politicians, most of them with little knowledge of the oil business, descended on the Middle East like a latter-day plague of near Biblical proportions. Few had the real measure of the problem; virtually all of them were present to serve their individual economies rather than the world economy. Apart from some little advantages here and there, the net effect of the flurry of activity was to underline heavily the power at the disposal of a cartel of oil producers. There was a scramble to acquire the small quantities of government crude, and at one stage the prices bid went to over $20 per barrel.

The oil companies have been accused of usurping the functions of governments during the crisis; the truth is that governments failed to face up to their responsibilities, and the oil companies had to fill the vacuum as best they could. While the main international companies did not foresee, any more than other observers, the precise outbreak of the Middle East war in 1973, they at least thought that there was a possibility of an oil shortage arising for one reason or another. They suggested in conversations with the main consuming governments that they should consult together to devise at least the bare bones of a rationing scheme to be brought into operation in the event of a shortage developing. In general, the main companies took the line that rationing of short supplies in an emergency was essentially a function for governments and not multi-national enterprises.

It was, of course, fully appreciated that any allocation scheme in times of shortage raises many difficulties. Is the base period to be historical or estimated future demand? Should indigenous production be brought into the picture or only oil traded internationally? If indigenous production of hydrocarbons is falling as in the United States, does this justify an increased allocation of imports? Should countries with a high *per capita* consumption, and therefore with a greater capacity to tighten belts, be cut more than countries with a low *per capita* consumption? These and many other problems of a rationing scheme can only be resolved eventually by a fair degree of arbitrariness. In the event, the consuming governments could not reach any agreement. They tended to concentrate on the inequities in the various proposals, with the inevitable paralysis of political will to do anything.

As the failure of consuming governments to grasp the nettle began to unfold itself, some oil-company spokesmen advocated at least a degree of restraint, since a scramble, in the face of a supply which was threatened to be reduced progressively every month, would result in a degree of chaos which would be in nobody's interest. The danger was recognized by several governments, but mutual suspicion and distrust dominated the field. Japan, for example, is particularly vulnerable because of its large imports of oil. In the absence of any sign of restraint elsewhere, the Japanese were afraid that the American Government would not be able to control the overseas purchases of their domestic companies. There was also a genuine fear, not wholly unfounded, that the home governments of the multi-nationals might bring pressure to bear on these enterprises to divert supplies to them at the expense of others. Through Japanese eyes, therefore, voluntary restraint could well have resulted in their suffering a greater cut than other countries. The end of the queue as a reward for virtue was unacceptable.

In their international operations the multi-national oil enterprises have a vast web of legal and moral obligations to the countries in which they operate and the customers with whom they do business. In the absence of any agreed international pro-rationing scheme, the only defensible action on the part of the companies was to apply the cuts as equitably as possible across the board. How far all international oil companies followed this route will only become known when the detailed history of the last few years is finally written. For the Royal Dutch/Shell Group with which the present author is connected, there was much debate but no doubt among its multi-national management about the course which should be followed. Of course it was necessary to comply with the destination controls imposed by the Arab Governments, but non-destination-controlled oil could be used to spread the load more equitably. The more intelligent Arabs realized that it was nonsense to think that institutions such as the Common Market could survive if Holland were completely deprived of oil while supplies were maintained to the surrounding countries. In any case the position of Rotterdam as the great entrepôt for the surrounding territories precluded any idea that Holland could be denied supplies.

It is easy to envisage the economic difficulties and the political acrimony that would have resulted if the major multi-national oil enterprises had succumbed to some of the pressures to which they were subjected. To some extent the same people who must share responsibility for the collapse of international leadership are now quite vocal about the need to have a code of international conduct for multi-national enterprises. A case can indeed be made for this; but the more urgent requirement is to devise a code of conduct for governments which will, as Professor Drucker put it, take account of the overall interest 'against all the partial and particular interests of its various members'. Following Dr Kissinger's speech in 1973 at the Pilgrims' dinner, some slow advance has been made in this direction. However, it would be unwise to over-estimate the degree of consensus achieved or to assume that chauvinism will play no role if the crunch comes again.

## Escalation of prices

The high prices paid in the scramble for the small quantities of government oil at the end of 1973 resulted in the OPEC Governments increasing their take yet again at the beginning of January 1974. Once more they regarded the small marginal quantities as reflecting the value of their crude oil. The progress of the increases in government take, in terms of a marker crude oil, are given in figure 4.1. This shows that in less than a year the government take increased by a factor of over four and a half.

The high prices also whetted the appetites of the producer governments for a speedier takeover of the oil concessions. As with the Teheran and Tripoli price agreements, the companies were informed that the participation agreements were dead. Some governments sought an immediate 60% while Saudi Arabia set its sights on a 100% takeover, with the old concessionaires providing the technical back-up and services in return for a fee. For a considerable time the companies knew neither the quantity of the oil that would be available nor the price that would be charged for whatever participation oil was sold back to them. In their inter-affiliate prices the companies averaged their widely varying costs for the two streams, but this did not prevent accusations of 'excess profits' although these accusations would only have been true if the companies had charged their affiliates the estimated buy-back price from governments. The pricing situation was near chaotic; figure 4.1 also shows the successive increases in government take that occurred during 1974 concluding with the 40 cents per barrel increase imposed in November by a Saudi Arabian-led splinter group from OPEC. It was a complex manoeuvre of reducing the posted prices, and therefore the buy-back prices, while imposing substantial increases in royalties and taxation rates. The higher government take which resulted was disguised as representing the 'return' (sic) to the consumer of the oil companies' 'excessive profits on the export of their crudes'. In fact the changes represented an additional transfer of funds of the order of $1·5 billion a year from the consuming countries to the three exporting countries of the 'splinter' group alone.

In the period after the second world war the bulk, but by no means all, of the investment decisions in the international oil business were made on an integrated basis. Starting off from an assessment of the possibilities in the individual markets world-wide where they did business; their relative

competitive strengths in each area and product; the attitudes of governments in such matters as supporting state enterprises, directly or indirectly;
and the investment intentions, as far as they were known, of competitors,
a profile was built up of the business possibilities open in the various parts
of the world. These were translated into investment requirements in retail
assets, depots, refineries, crude and product tankers, production facilities,

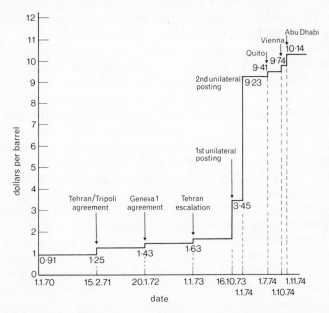

**Figure 4.1**   Producer government revenue marker crude (Arabian light 34°) free on board
Ras Tanura. (1) Up to end 1972 calculated as Host Government Take on companies' equity
crude. (2) From 1.1.73 calculated as total Producer Government Revenue.

and harbour and pipeline requirements. As is fairly common in an
expanding capital-intensive industry, the totality of projects which
individually met the investment criteria sometimes proved too great for
the financial resources—depreciation, retained profits and the cash that
can be raised in the market place by borrowing on rights issues—of
individual groups, and choices had to be made. In spite of this, the oil
industry, because of its flexibility, was able to meet the demands placed
upon it, and indeed filled the energy gaps which resulted when coal and
nuclear power failed on several occasions to meet the targets which had
been set. As a result of the uncertainties created by recent events, the basis
of the integrated investment decision-making has been disrupted without
any very clear alternative taking its place. The full effect of the change

has yet to be felt, since initially the Arab cutbacks and subsequently the combination of high crude-oil prices and a fairly general economic recession, has resulted in a fall in demand which in turn has thrown up considerable surpluses in all facets of the business—crude-oil production, tankers, refineries and distribution assets. Logistic disruptions due to uncertainties and delays have therefore been relatively easily accommodated. This should not be allowed to disguise the major question as to whether the investment required to ensure future energy supplies will be forthcoming, a problem which will be examined later.

Now few people in the industry really expected that the cheap oil of the sixties would last much beyond the middle eighties. Indeed, if market forces had been left to determine prices, they would have expected gradual, but not precipitate, rises some time within this period and certainly towards the end of it. Even before the crisis broke, there had been a worldwide trend towards offshore exploration where, in the event of success, the technical costs of production are many times greater than the low level prevalent in the main Middle East fields. As offshore exploration progressed into deeper waters, costs would have continued to escalate until a cross-over point was reached where coal would become progressively more competitive with oil, and atomic energy would tend to pre-empt the primary energy requirements of base-load power stations. Eventually cost levels of conventional crude oil would be such that tar sands and shale, particularly those in Canada and the United States of America, would be developed. Since the world reserves of coal, tar sands and shale are many times greater than those of conventional oil, this scenario would have seen the world well into the twentieth century. There can also be little doubt that gradual adjustments to rising prices and costs of energy could well have been accommodated by the flexibility of the economies of the Free World. By exercising their near monopoly powers, the OPEC Governments have telescoped the time span, and both the suddenness and the magnitude of the price increases are imposing considerable strains on the international monetary and trading systems. The action which will be required to avoid a breakdown will, to some extent, depend on whether present high prices are likely to last for only a brief or for a longer period.

## Will OPEC survive?

At any particular time a cartel of producers of almost any commodity can force up the price by reducing supplies. The degree of success will largely depend on how essential the commodity is, the availability of substitutes, and the lead-in time necessary to develop alternative sources of supply. Economic history shows several examples of the collapse of cartels, particularly when prices are escalated to many times the technical cost of production. This is certainly the case with oil and Professor Adelman* (1972) is convinced that the OPEC cartel will collapse just as surely as the Stevenson Rubber Restriction Agreement did in the inter-war period. Adelman starts off his argument from the correct premise that there are abundant reserves and no real shortage of conventional oil; nor indeed is there likely to be any shortage until into the middle eighties. He claims that prices have been kept artificially high by what he asserts to be a producer government/oil company cartel, and once the latter are squeezed out in whole or in part from the crude-oil production side of the business, the governments as independent crude-oil suppliers will compete vigourously for the available outlets, thus bringing prices tumbling from their present levels. But Adelman's analysis is defective on several counts. There never was, and there is no evidence to suggest that there was, a producer government/oil company cartel. Even a superficial study of the events in Libya over the last couple of years provides considerable evidence to the contrary. What we witnessed in the seventies was a build-up of pressures on the companies by the host governments; and the companies, in the absence of any countervailing power, had little alternative but to accept or have their assets nationalized at no, or derisory, compensation. The extremists set the pace and once acceptance of their terms had become fairly general within their own country, it was only a matter of time until the terms spread to other areas. It is also quite wrong to suggest that the main oil groups were not fully aware that additional host-government taxes would represent burdens on the balances of payments of the consuming countries. The governments of these countries were kept informed of developments; some evaded comment, but others took the line that continuity of supply must take precedence over cost. Moreover, the producer government/company cartel theory of Adelman is hardly consistent with the admitted historical record that the massive increase in producer-government take imposed in January 1974 was done without any form of consultation with the companies.

Even if the companies were forced out completely from crude-oil production, it is difficult to see how this fact alone should make the producing governments behave differently than they have been doing on the matter of prices. The producing countries are not, of course, all subject to the same economic pressures. Some of them—Iran, Iraq, Nigeria and Indonesia, for example—have large populations, ambitious development programmes, and therefore a considerable absorptive capacity for imported goods and services. Even so, the magnitude and speed of the increases in government revenue when taken together with the time needed

* Professor of Economics, Massachusetts Institute of Technology.

to mount development schemes, have resulted in these countries building up surplus funds; but in these cases it is likely to be a temporary phenomenon. At the other extreme some of the remaining big producers—Saudi Arabia, Kuwait, Abu Dhabi and Libya, for example—have small populations, and therefore a relatively small absorptive capacity for imports of goods and services; their build-up of funds is likely to be rapid and cumulative.

With OPEC revenues as a whole running currently at an annual rate of some US $60/70 billion above absorptive capacity, it is difficult to see how there can be an overall pressure to intensify inter-governmental competition in the crude-oil market at the expense of price. Saudi Arabia has the greatest surplus revenue and is also the OPEC member with the largest production and reserves. If one of the other OPEC governments, with a large population, were to increase its production, there would be no hardship in Saudi Arabia closing in a corresponding quantity if necessary to maintain prices. Indeed this could take place in any case if the Saudis fail to find sufficient investment outlets for their surplus funds. As Sheikh Yamani put it, there could well come a time when the value of a barrel of oil in the ground exceeds another Treasury Bill in London or New York.

If crude-oil reserves and populations had been more evenly distributed among the producing countries, Professor Adelman's assessment of probable future trends would be more sustainable, but still unlikely. However, the underlying facts that some of the major reserves are concentrated in areas of low population with no very perceptible pressure to increase income tends to destroy Adelman's rather 'textbook' prognostication of the nemesis of OPEC. This does not mean that prices may not be reduced in the next five to ten years; they could well be, but if they are, it will probably be the result of political decisions and not the operation of the traditional economic forces that have undermined cartels in the past.

### What is a fair price?

There has been considerable discussion about what constitutes a 'fair price' for crude oil; like justice and equity, fairness is a happy hunting ground for politicians. On a market basis the price would approximate to the cost of producing the marginal barrel and, since costs vary widely, considerable economic rents would accrue to the lowest-cost producers. The problem is complicated in the oil industry, at least as far as the major companies are concerned, since there always has been a latent conflict between cheapness and security of supply. The former pulls in the direction of concentrating on the lowest-cost producers; the latter points to diversity of supply sources, and therefore production at varying costs. A balance, albeit not an easy one, was struck in this conflict by each major oil company according to its particular circumstances. However, once we depart from costs as being the main ingredient in the build-up of prices, the discussion usually becomes bogged down in a morass of subjective and, only too often, emotive judgments. This is particularly so when dealing with a finite resource which is subject to depletion, as compared with an annual crop such as wheat or soya beans. Provision must be made against the day when the resource is finally exhausted; the whole concept of royalties, with its charge per unit of oil, coal, or ore produced, was designed with this depletion problem in mind. But how much should the royalties be? In the case of crude oil it could be argued that where technical costs are in the range of 10–20 cents per barrel produced, a combined royalty and government tax of $1 per barrel seems generous; but it could be equally argued that the figure should be at least $8, since that could well be the cost of producing the alternative in a decade's time. There is in fact no objective means of ascertaining a 'fair price' when the market no longer operates. But one thing is clear—a virtual monopoly in a product such as crude oil, with no readily available alternatives and produced in the circumstances already described, enables the monopolists, if they are so inclined, to charge today at least the price they estimate will prevail five to ten years ahead if normal market forces were to reassert themselves.

### How important is North Sea oil?

Considerable attention has been focused on the oil finds in Alaska and the North Sea as providing some sort of counter-balance to OPEC domination of the energy scene; but while important for the countries directly concerned—the United States, Britain and Norway—they are relatively small in a world context. Table 4.4 sets out the Free World's reserves of conventional crude oil.

Owing to the long lead-in time for the development of coal mining, tar sands, shales and nuclear power, there is no really credible alternative to these conventional reserves as the principal component of the Free World's energy supply over the next decade. And with the bulk of the reserves concentrated in the Middle East there is no credible alternative to Middle East Oil.

### Recycling oil revenue

Against this background certain actions are necessary on the part of the consuming governments. The most immediate problem is to prevent the substantial accumulation of financial resources in the hands of OPEC Governments from disrupting the international financial and trading mechanisms. The estimated OPEC 'surplus' for 1974 of some $60–$70 billion is likely to grow cumulatively over the years as a result not only of additional oil sales but also as a result of interest earned on the balances. By the end of the decade these surplus funds could well be in the range of $400–$650 billion. They are, of course, the obverse of the current balances of payments deficits of the oil-importing countries. The developing countries, particularly a country like India, have been very hard hit. On an unrestricted basis the high cost of oil imports in 1974 would have absorbed around half of India's export earnings. In absolute terms, however, the main drain falls on the industrialized countries of Japan, Europe and the United States. Yet more than a year after the crisis hit these countries, there is still no international mechanism which ensures

**Table 4.4**    Proven reserves of crude oil for the World outside USSR, Eastern Europe and China as at end of year 1973 (*World Oil*, 15.8.74)

|  |  | (*Billions* ($10^9$) *of barrels*) |
|---|---|---|
| MIDDLE EAST |  | 316 |
| of which: | Saudi Arabia | 97 |
|  | Kuwait | 73 |
|  | Iran | 68 |
|  | Iraq | 36 |
| AFRICA |  | 57 |
| of which: | Libya | 23 |
|  | Nigeria | 18 |
|  | Algeria | 10 |
| NORTH AMERICA |  | 47 |
| of which: | USA | 35 |
|  | Canada | 9 |
| CARIBBEAN & SOUTH AMERICA |  | 26 |
| of which: | Venezuela | 14 |
| FAR EAST & AUSTRALASIA |  | 19 |
| of which: | Indonesia | 12 |
| WESTERN EUROPE |  | 17 |
| of which: | UK | 11 |
|  | Norway | 5 |
| WORLD OUTSIDE USSR, EASTERN EUROPE & CHINA |  | 482 |

that the surplus funds are recycled back to the countries from which they were obtained pro rata to their requirements.

It is not that there is any absence of ideas as to how this potentially dangerous situation should be handled. The Bank for International Settlements devised what could have been a virtually automatic mechanism for crediting and debiting the producing and consuming countries in respect of their oil transactions. The scheme had the additional merit of minimizing the enormous, and possibly disruptive, movements across the various foreign exchange markets. Subsequently there were other proposals by Messrs Witteveen, Healey and Kissinger. Some of the schemes required the cooperation of at least a few of the major crude-oil producers and some of the major consuming countries. At least one of the schemes could be handled by the major industrial countries themselves since, once the oil is sold, the proceeds received by the exporting countries are within a closed financial system of America, Europe and Japan. True the OPEC Governments could, within limits, switch from one country to another within the system, but this is expensive and its adverse effects can be minimized by appropriate swap agreements among the central banks concerned.

There are arguments for and against the various schemes—just as there were arguments for and against the various international oil-rationing schemes in October 1973. The main requirement is to obtain a consensus on one or more of the proposals—they are not mutually exclusive—even although the outcome may fall short of perfection. The money markets of the world have handled the problem for over a year; but with the growing magnitude of the sums involved, the tendency for these to be deposited short-term when the deficit countries need to borrow long-term, and the fact that some of the deficit countries in most urgent need are the least credit-worthy, set limits to the amounts that can be handled in the traditional markets.

The real risk that we run is that the progress which has been effected since the war towards a more liberal international order in finance and trade will be arrested or reversed. In the absence of some form of international agreement about the recycling of the surplus oil revenues, there is a temptation for some governments to endeavour to achieve balance in their international current accounts by artificially stimulating exports, restricting imports and the other devices which were so familiar in the inter-war period. But since the absorptive capacity of the oil-exporting countries as a whole is limited, such a policy, by any major country, or group of countries, must be self-defeating for the world as a whole, since it will only aggravate the current balances-of-payments problems of the remaining governments, who may in turn retaliate. To write of the possibilities of international deflation at the present moment might, to reverse John Maynard Keynes' aphorism, sound like warning a man who is suffering from corpulence of the dangers of emaciation; but if international leadership founders once again, the danger is certainly there.

It is also essential to effect greater economies in the use of energy along the lines which have been developed recently at length. It is not possible to change the profile of the motor-car population overnight; but the sharp increases in prices are likely to result over a period in a greater concentration in lower-fuel-consumption vehicles. More intensive insulation of factories and homes, which may have appeared marginal investments in the sixties, show a rapid pay-out in the vastly changed circumstances of today. The policy of holding prices of the various forms of energy down to artificially low levels should be reviewed. The low price for natural gas in the United States stimulated consumption and inhibited the search for additional supplies; indigenous production is now falling appreciably short of requirements. In the United Kingdom, virtually the whole range of energy supply, but particularly the products of the nationalized industries, were kept below the market level. By way of illustration, electricity is one of the most convenient forms of energy for many uses, but it has a relatively low efficiency as a form of heating. Although subject to constant improvement, net generating efficiency is of the order of 33%, while around 12% is lost in transmission from power station to consumer. With a space-heating effectiveness of some 95%, the overall efficiency is about 28%, compared with around 60% for heating by oil or gas. To hold electricity prices down artificially is merely to distort the consumption pattern and increase inefficiency in use.

**Alternative sources of energy**

The Free World must also develop other sources of energy supply as quickly as is economically feasible. However certain impediments, not anticipated a few years ago, are now appearing on the scene. It has for long been recognized that alternatives to Middle East oil would be more costly to produce and considerably more capital-intensive; what was not anticipated is the degree of inflation which has been much more severe in capital equipment than in the general range of consumer goods. When the first exploration successes were achieved in the North Sea, it was then estimated that the investment cost per barrel per day would be of the order of US $2500 or approximately ten times the Middle East level; current estimates for future development costs are between US $5000 and US $7000 per barrel per day. Broadly the same escalation is taking place in the investment costs for producing oil from tar sands. The capital requirements estimated some two years or so ago were in the US $8000–$10,000 range per barrel per day; the figures are now in the area of US $15,000 and above. Sharply escalating costs have also hit atomic power station programmes, and the uncertainty created has cast doubts on the implementation of certain projects which had looked virtually certain a year or so ago.

The second impediment which is appearing is the restrictive attitudes of various governments. Exploration for oil has always contained an appreciable gambling element. Several tens of millions of dollars can be spent, and the results may vary between zero and a find of such dimensions as will show a large return on the capital to be invested in development. The lure of the bonanza motivates the oil explorer just as it does the 'investor' in the football pools. The moralists in their cloisters may deplore it; but it is a fact of life with which the industry has learned to live. Now the pools' 'investor' knows that if his successes are heavily taxed and he is left to carry his losses, the attractions of filling in the weekly coupon rapidly diminish to vanishing point; and the same is true of the search for crude oil. If governments become obsessed, as some of them are, with the hole instead of the doughnut, they should not be too surprised if they finish up with many holes and rather few doughnuts. Indeed, with some of the present restrictive attitudes, it is becoming doubtful whether the industry will be able to attract the vast amount of capital that is necessary to create a credible alternative to Middle East oil.

The third impediment is the threat of nationalization in whole or in part on unspecified terms. As already mentioned, net book value, which has been enforced by some of the OPEC Governments, could not be regarded as in any way representing adequate and effective compensation. The constant threat creates considerable uncertainty, and uncertainty on major issues of this kind has an inhibiting effect on investment. In Britain the Socialist Government has stated its intention to 'negotiate' a 51% participation in all successful North Sea ventures. There is a fair degree of intellectual confusion as to what the Government can achieve by virtue of its position as a government, and what it can do in its capacity as shareholder. In its capacity as a government it can, by means of taxation, obtain a stake in the future profitability of the North Sea; it can, by legislation, enforce depletion rates and so forth; it can do none of these things by virtue of being a shareholder. Indeed it is largely a myth that owning 51% of a company's shares gives control. The law does not look kindly on directors who abandon their fiduciary responsibilities to look after the interests of the company as a whole, in favour of pursuing the interests of one group of shareholders only. What is the government's aim in seeking 51%? Not to put in management talent; it is no secret that they dispose of no surplus of this commodity. Not to contribute expertise; they dispose of no surplus in that commodity either. It is difficult to escape the conclusion that it is but a profound obeisance to doctrinaire socialism, and it is one that will certainly not hasten the day when North Sea oil will ease Britain's balance of payments' position.

The analogy with events in the Middle East is quite misleading. There, only a few companies are involved, and they are functioning on a production basis. The North Sea has a multiplicity of companies and is still very much at the development stage. As would be expected, the companies vary in ability and in their engineering solutions to some of the complex problems involved. Inevitably mistakes are going to be made. Either the Government provides its 51% of capital requirements and relies entirely on the technical managements of the various companies, or it builds up a staff to make its own checks on the many budget programmes. Yet the staff the proposed British National Oil Corporation requires is precisely the sort of staff which is in very short supply. To the extent that the Government succeeds in recruiting, it is likely to slow up existing work in the North Sea; to the extent that it fails, vast sums of public money— well over £2000 m—could be disbursed without that degree of supervision the Public Accounts Committee would regard as adequate. The dilemma is real.

The fourth hurdle which could slow up the development of alternative energy sources is the ecological lobby. The location of atomic power stations, sites for drilling platforms, routes for pipelines, open-cast-mining permits and so forth, all require planning permission; and the views of the ecologists must be heard. The rights and wrongs of the particular issues are beside the point. The procedure slows down the development of alternative energy supplies; and somewhere along the line there has to be a compromise between the two requirements of the community.

As was shown in Table 4.1, the primary energy demand of the world outside the communist areas totalled 86·9 million barrels a day of oil equivalent in 1973. Of this total, some 56% was supplied by oil; coal and natural gas supplied 18% each, while hydro-electricity and nuclear power supplied 7% and 1% respectively. Table 4.5 shows a range of possibilities— depending on the assumptions made about economic growth and availability—for the year 1980.

Dealing with the non-oil sources first, nuclear power shows the most spectacular increase and, in some respects, is the one which shows the least degree of uncertainty. To be on stream by 1980 the plant must already be on order. The doubts that can arise about nuclear power's contribution is whether the plants on order can be produced on the agreed schedules, and the performance of the plants when they go critical. By the end of

1980 it is estimated that the free world's nuclear capacity will be approximately 250,000 megawatts, of which 55% will be in North America, 31% in Western Europe, 11% in Japan, and the balance in other areas.

Table 4.5 also shows a considerable improvement in coal. With reserves greater than all other fossil fuels combined, and with these distributed more evenly than oil around the world, the improvement in relative position is likely to continue in the period beyond 1980. Although coal

**Table 4.5** Primary commercial energy demand/supply in the world outside the communist areas for 1973 and 1980

*(Million barrels per day oil equivalent)*

|  | 1973 (Actual) | 1980 High | Low |
|---|---|---|---|
| Oil | 48·4 | 65·0 | 55·5 |
| Coal | 15·7 | 21·6 | 19·8 |
| Natural gas | 16·1 | 18·4 | 18·0 |
| Hydro-electric | 5·8 | 7·0 | 7·0 |
| Nuclear | 0·9 | 6·4 | 6·2 |
| Total | 86·9 | 118·4 | 106·5 |

has a very developed production technology, both in deep mining and open-cast, it is still labour-intensive and costs are sensitive to wage rates. Any spectacular increase in coal's contribution will require considerable development of strip mining, which not only produces the understandable ecological problems already mentioned, but also raises difficulties similar to the Athabasca tar sands in the way of manufacturing excavating machines of very large capacity—the lead-in time for a major dragline is of the order of four years.

If coal is to penetrate the energy market on a large scale, it will not only be as a source of energy for steam raising, but also by way of conversion either into low or high calorific-value gas, or by liquefaction to replace conventional refined oil products. The basic techniques for conversion are known, but they suffer from high capital costs and low efficiences of between 60% and 70% compared with around 93% for crude oil. With inflation at current high but uneven rates, it is difficult to assess competitive costs. The best estimates available to Shell would indicate that in North West Europe and Japan coal (after allowing for diseconomies in use) would be competitive for steam raising at an f.o.b. crude oil cost in the Persian Gulf of $6–$9 per barrel. As a synthetic oil produced at the coal source, it would be competitive with a 'free on board' crude oil cost in the Persian Gulf of $10–$14 per barrel.

128 ENERGY—THE INTERNATIONAL SCENE

Table 4.5 shows oil as having the largest absolute increase up to 1980, but the average annual increase—at 2% on the lower estimate and 4·3% on the higher—is well below the trend line since the end of the war. Energy is undoubtedly available to meet the world's requirements; the uncertainty is whether governments will permit its production, and on what terms.

In the much longer run scientists assure us there is no need for concern, since energy will ultimately be produced from resources such as nuclear fusion and the sun, resources which are virtually infinite. No doubt this is true, but the problems are formidable, and the journey from the laboratory via the drawing board to everyday use can be a long one; and before we reach this age of plenty we have to ensure survival. To do so in a reasonable degree of comfort will require more statesmanship on the part of politicians than we have witnessed in recent years.

## FURTHER READING

Adelman, M. A. (1972), *World Petroleum Market*, Baltimore and London, John Hopkins University Press.

Adelman, M. A. (1972), 'Is the Oil Shortage Real?', *Foreign Policy*, New York, Foreign Affairs Inc., no. 9.

Drucker, P. F. (1969), *The Age of Discontinuity*, Heinemann.

CHAPTER FIVE

# THE ATOM AND THE ENVIRONMENT

H. J. DUNSTER

## Introduction

As a species man has developed in the presence of, and probably in part because of, environmental radioactivity, but only in the last generation has he been able to add an artificial component to this natural background. For a variety of reasons the public has not always found it easy to come to terms with the presence of radioactivity in the environment. Perhaps the main reason for the difficulty has been the precipitate rate at which sources of artificial radioactivity have been introduced, and the political and social implications of some of these sources.

For most people, their first recognition of environmental radioactivity must have come from the publicity about the fallout of debris from the testing of nuclear weapons. To a natural fear of the unknown were added feelings of guilt or, in some countries, of envy, and the sharper fear of a new and catastrophic war. Much of this natural alarm has been transferred to the developing nuclear power industry and has also found a place, not always healthily, in the wider movement to maintain and protect our environment.

Now, some 30 years after the first fallout from a nuclear weapon, and 20 years after the start of the nuclear power industry, it seems an appropriate time to re-examine the place of the atom in our environment.

## Sources of radiation in the environment

Sources of ionizing radiation in our environment may be natural or artificial and, although any consequences they may have will be similar in kind, it is often appropriate to consider them separately.

129

*Natural sources*

The everyday background of naturally occurring radiation is made up of several different components. Probably the best known is *cosmic radiation*. This consists of particles of very high energy, some originating from the sun and the others from even greater distances. Almost all these particles are stopped by collisions with gas molecules in the upper layers of the atmosphere. This process causes the ionization which makes the electrical properties of our upper atmosphere so complex. Phenomena like the reflection of radio waves and the aurora borealis (the northern lights) are attributable to this ionization. Some of the particles collide with atomic nuclei, producing secondary particles, including neutrons. These processes also produce substantial amounts of tritium and carbon-14, the radioactive isotopes of hydrogen and carbon. Eventually a complex mixture of particles reaches the earth's surface. The intensity of this flow of particles is higher at greater altitudes above sea-level because of the reduced atmospheric shielding at these altitudes.

The second natural source of radiation in our environment is the *natural radioactivity* of the earth's crust and the atmosphere. Elements with atomic numbers above 82 (lead) are all radioactive, and there are complicated chains of radioactive decay from the heaviest natural element, uranium, down to lead. In addition, there are a few isotopes of lighter elements which are naturally radioactive and which have half-lives long enough for them to have survived from primordial times. Potassium-40 is the most important of these. Finally, there are tritium and carbon-14, produced in the upper atmosphere by cosmic rays. This range of materials is so wide that it can be said that any common material is slightly radioactive. Living matter, including our own bodies, contains hydrogen, carbon and potassium; rocks, soil and building materials, all contain small amounts of potassium and of uranium and its daughter products, and thus all are radioactive to some extent.

It is important to obtain a quantitative assessment of the amount of radiation to which man is exposed from these natural sources, because this forms one basis for judging the implications of man-made additions. The most widely used quantity for expressing the 'amount' of ionizing radiation is the *absorbed dose*, which is a measure of the amount of energy absorbed per unit mass of a material, including living tissue. The current unit of absorbed dose is called the *rad*, equal to the absorption of 0·01 joule/kilogram of matter (J/kg). The rad is a fairly large unit and the millirad ($\frac{1}{1000}$ rad) is often more convenient. The rad will eventually give way to a new unit, the *gray*, equal to 1 J/kg, to bring it into line with the International System of Units.

Table 5.1 shows typical natural contributions to the annual dose to an inhabitant of the British Isles. The cosmic-ray component is enhanced for those who live well above sea-level by about 50% at 1500 metres (5000 feet), and the dose from the earth's crust is higher in areas of high natural radioactivity, commonly by a factor of about 2 from the average. Most people in this country receive naturally doses somewhere between 80 and 150 millirads/year. Part of this variation is due to our own actions—choosing where to live and using bricks or stone as building materials in place of the slightly less radioactive material, wood. Much wider variations in natural dose can occur as a result of some specialist occupations. Thus, airline crews are exposed to rather more cosmic radiation, and miners to more radiation from rocks than is the average person.

Table 5.1 Annual radiation doses from natural sources in the United Kingdom

| Source | Typical annual dose (millirads/year) |
| --- | --- |
| Cosmic radiation | 30 |
| Environmental radioactivity | 40 |
| Radioactivity in the body | 20 |
| Total | 90 |

*Artificial sources*
The largest man-made source of radiation exposure is not really environmental, although it affects almost everyone at sometime or other. It is the use of X-rays and radioactive materials for medical diagnosis and treatment. The doses are to limited parts of the body and vary very widely from one individual to another, so it is difficult to make useful comparisons with the more uniform dose from natural background. One basis of such a comparison is the so-called 'genetically significant dose', which is a measure of the possible harm passed on to our descendants. It can be thought of as a measure of the average dose to the population. On this basis, medical exposure, including surgery, dentistry and so on, amounts to about 20% of the average natural background.

Since the 1950s, the most widespread contributor to artificial radioactivity in the environment has been the debris from nuclear explosions. Test explosions in the atmosphere release large quantities of radioactive material, most of which is lifted into the stratosphere by the heat of the explosion. The material subsequently falls back to earth and causes radiation dose to man by direct exposure to the deposited activity, and by becoming transferred into the body via food, especially milk. The fallout contribution to man's dose has been declining for some years now, despite the occasional atmospheric tests that still occur. The total contribution to the dose from all the tests up to 1974, allowing for future contributions from material not yet deposited, is about the same as the dose from 2 years' exposure to natural background. The current dose rate from the material deposited on the ground and in our bodies is about 2 millirads/year in Britain, or about 2% of the dose rate from natural background.

The number of artificial radiation sources in our everyday lives is surprisingly large. Most of these are luminous devices of various kinds— clocks, watches, compasses, and even telephone dials—and the radiation dose they deliver is very small. Radiation in the form of X-rays from television sets has sometimes received publicity; considerable care is taken in the design to keep the emissions to a minimum and to ensure compliance with the now widely accepted international recommendations. Together these miscellaneous sources of radiation contribute an average dose to the population of rather less than 1% of that from natural background.

Finally, there is the presence of artificial radioactivity in our environment resulting from the use of radioactive material in industry, medicine, research and teaching, and the inevitable production and processing of radioactive material in the nuclear industry. All these processes give rise to some radioactive waste, and it is with these wastes and their implications, particularly those from the nuclear industry, that this chapter is principally concerned.

### The significance of radiation doses

Ionizing radiation, by definition, changes atoms and molecules by removing electrons from the atomic structure. Individual atoms rapidly regain an electron and thus repair themselves. Molecules may do the same, but may not reform in quite their original structure. Some of the damage is then not, or not immediately, repaired. If these changed or damaged molecules are in living cells and are sufficiently numerous, the cell may be damaged in a way which may or may not be repairable. If the change is severe enough, the cell may be killed. If enough cells are damaged or killed, the health of the whole organism may be jeopardized. Since the transmission of information from one generation of cells to the next, and from one generation of man to the next, depends on the replication of molecular structures, the possibility of faulty replication of cells and of genetic, i.e. inherited, defects in man resulting from ionizing radiation is a real one. The chain of events from ionization to injury is long and involved, and there is a complex interaction of the processes of damage and repair at all stages, even in the final process of replication of cells. There is thus no simple answer to the question: how dangerous is radiation? However, some facts about the effects on man are established with certainty, and others can be inferred with reasonable certainty from the known effects in other animals.

The following effects in man are known beyond all reasonable doubt to occur following doses of radiation delivered in short periods.

High doses (500–1000 rads and above) to the whole body are fatal.

Even higher doses (1000 rads and above) to local parts of the body can produce serious, even irreparable, tissue damage.

These high doses, and lower ones down to about 50 rads, cause an increased risk of eventual induction of cancer.

Still lower doses (a few rads) to the foetus at critical times can produce developmental injuries.

The following further effects may well occur, but have not been conclusively demonstrated in man and cannot be inferred with certainty from experiments in animals.

Spreading the dose over long periods probably reduces the risk of long-delayed serious effects such as cancer (it certainly reduces substantially the likelihood and severity of acute effects).

Doses lower than 50 rads in a short time may also produce a higher than natural risk of cancer, but the relationship between dose and risk is not known.

Doses of a few rads to the foetus probably increase somewhat the normally low risk of cancer in the first 10 years of life.

Finally, here are a few statements on topics which are the subject of controversy because adequate facts are not available.

The additional risk of cancer resulting from a radiation exposure may be proportional to dose at any level, however small.

The artificial reduction of dose below the normal level of natural background may have deleterious effects on health, at least for some organisms.

Small doses above the level of natural background may improve the biological performance of some organisms and may cause some prolongation of life.

These last two statements are not necessarily incompatible with the first, since radiation may, at the same time, be doing both good and harm, but they are usually regarded as being highly unlikely. The statement about a direct proportionality between dose and risk, however, is usually taken as a working hypothesis in establishing the basis of radiation safety. It is important to remember that it is a hypothesis and not an established fact.

*Radiation protection standards*

Primary radiation protection standards are aimed at limiting both the short-term and the long-term radiation doses to the exposed individual, so that he is not at significantly greater risk than other people. Consequently, we cannot demonstrate any harm to health at these levels of exposure; and we have to predict the harm, if any, from our now considerable knowledge of the long-term effects of much larger doses delivered in much shorter times. Since the mid-1950s, these predictions have been made on the very cautious hypothesis that the probability of delayed serious effects, such as cancer, is proportional to the absorbed dose and is unaffected by the time over which the dose is spread, whether it be a few seconds or a lifetime. In recent years, the last part of this hypothesis has become increasingly unlikely, and some allowance is now being considered for the reduced effects of protracted doses.

Because of the possibility that even small doses of radiation may carry some increased risk, the protection standards include a requirement to keep doses as far below the permitted limits as can reasonably be achieved. In judging what is reasonable, account has to be taken of the economic and social costs of achieving further reductions. There is a net loss to society if the effort applied to reducing an already small risk could have been better employed in some other more effective way. It is therefore not necessarily sound policy to press for lower and lower radiation exposures.

## Environmental limits

The primary protection standards are more easily applied in the control of environmental radioactivity if they are used to provide a basis for secondary environmental limits, such as those applied to the concentration of radioactive materials in foodstuffs, or to the amount of radioactive waste that may be released to the environment. In the United Kingdom, the main emphasis has been put on the release limits, which can be enforced much more effectively than can limits applying to doses to individual members of the public. The underlying aim, however, is to achieve an adequate and appropriate standard of control over the doses to individual members of the public, and to the population as a whole. The environmental limits are thus tools rather than standards in their own right.

## The nuclear-energy industry

The nuclear-energy industry is often thought of as a self-contained entity, but it is better seen as a part of a much larger complex of industries involved in the supply and distribution of energy. In particular, the decision to use nuclear energy is not an isolated decision—it is a choice between one sort of energy and another. Such a choice involves the weighting of many factors, economic, social and environmental, for all the available alternatives.

## Nuclear fission

The present nuclear-energy programmes are based on the process of fission in uranium and, to a small extent, plutonium. If the uranium isotope, uranium-235, is bombarded by neutrons, its nucleus splits into two roughly equal parts, fission products, and several neutrons. These neutrons can be made to cause fission in further atoms of uranium-235 and thus maintain the process. Some of the neutrons react with the more common isotope of uranium, uranium-238, and this reaction leads to the formation of the plutonium isotope, plutonium-239, which then begins to play a part in the chain reaction in much the same way as the original uranium-235. The fission products from both uranium-235 and plutonium-239 are emitted with considerable energy, which they transfer by collision to the atoms and molecules making up the fuel, and which thus appears eventually as heat.

The nuclear reactor consists of an array of uranium fuel in a configuration that allows the fission process to continue, and to be controlled by the deliberate absorption of any excess neutrons. A coolant is passed through the fuel assembly, usually called the *core*, to transfer the heat to a boiler where steam is raised and used as in a conventional power station. In some types of reactor, known as *boiling-water reactors* or *steam-generating reactors*, the steam is raised within the core, and there is no separate boiler.

Because the loss of uranium-235 is not fully compensated by the gain of plutonium-239, and because the fission products interfere with the mechanical and nuclear properties of the fuel, the irradiated uranium cannot be used indefinitely. Its life in the reactor is long, typically several years, but it has then to be replaced by new fuel. The irradiated fuel, sometimes erroneously called *spent fuel*, contains residual uranium, now depleted in the fissile isotope, uranium-235, the fission products, which are radioactive isotopes of elements in the middle of the range of atomic number, roughly from $Z = 30$ (zinc) to $Z = 63$ (europium), and plutonium. Because of multiple nuclear processes in the reactor, the plutonium consists of several isotopes in addition to plutonium-239 and is accompanied by materials of even higher atomic number, such as americium ($Z = 95$) and curium ($Z = 96$). The depleted uranium and the plutonium are potentially valuable materials for recovery and use in new fuel, while the fission products and other elements are waste materials.

The principal stages of the so-called *nuclear-fuel cycle* can now be identified. They are illustrated in figure 5.1. Uranium is injected into the cycle from the mining and concentration of ores. For most types of reactor the uranium has to be slightly enriched in uranium-235, and the enriched uranium is then fabricated into fuel, used in the reactors, and sent to a reprocessing plant. Here the irradiated fuel is chemically processed to separate it into its main components, the uranium, the plutonium and the fission products. The recovered uranium is sent for re-enrichment and for recycling through the reactors. The plutonium is at present stored, but there are plans for it to be used, together with some uranium, as a fuel for the so-called *fast breeder reactors*. These reactors make more effective use of the neutrons released in the fission process, and can produce more plutonium than they consume by irradiating a uranium blanket surrounding the core. The blanket can be made of uranium depleted in uranium-235, because it is the heavier and commoner uranium-238 that is the source

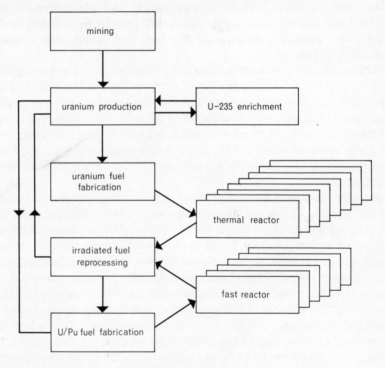

**Figure 5.1**    The nuclear-fuel cycle.

material for more plutonium. Depleted uranium is a waste product from the enrichment plant, and its utilization in fast reactors increases the amount of energy that can be extracted from a given amount of uranium fuel by a factor of about 50.

The introduction of the fast breeder reactors adds a further loop to the fuel cycle, with plutonium being extracted from both the partly used core and the uranium blanket, and being reprocessed to give fuel for further reactors. By the time there are enough fast reactors to provide a substantial part of the nuclear-energy programme, there will be a substantial stockpile of depleted uranium, and the requirement for new supplies of uranium will be dramatically reduced.

### Nuclear fusion

The possible alternative to the fission process is the *fusion* of light atoms. If the nuclei of light atoms, notably isotopes of hydrogen, can be made to collide with sufficient energy, they combine to form heavier nuclei and neutrons. The process is accompanied by a release of energy. This process is easily accomplished by accelerating individual nuclei by various electrical means, but these consume far more power than is released by the nuclear reactions. An alternative method is to heat the light elements to very high temperatures. The individual nuclei then reach such high speeds that successful fusion collisions can occur. If the elements are also kept under very high pressure, these fusion collisions become frequent, and there is a massive release of energy. This process, too, is feasible, but at present only in the conditions existing in the centre of a nuclear explosion. It is the basis of the so-called *hydrogen bomb*, where a fusion reaction is detonated by a fission explosion.

Very high temperatures can also be achieved by containing and compressing the light elements within a complicated combination of magnetic fields. The aim of current developments is to maintain the light elements in this magnetic containment at a sufficiently high temperature and for a sufficiently long period to allow the reactions to release more energy than is needed to operate the magnetic fields. Considerable progress is still needed before it can be said that controlled fusion is an alternative to nuclear fission.

### The environmental consequences of nuclear energy

Nuclear fission and, if it is successfully developed, nuclear fusion both produce very large quantities of radioactive materials. It is these that give rise to the possible adverse environmental effects of the nuclear generation of energy. Almost all the radioactive material is kept within the fuel cycle, the uranium and plutonium as fuels, and the fission products as a waste which is stored while it decreases by radioactive decay. However, no process is capable of achieving total containment, and there is some inevitable release of radioactive materials to the environment at various stages of the nuclear fuel cycle. The possibility of unplanned releases in accidents to nuclear reactors and to chemical engineering plants cannot be totally excluded, though it can be made very small. These two aspects of nuclear energy—the release of radioactive waste and the possibility of serious accidents—have become major centres of controversy round the nuclear-energy industry.

## Waste management in the nuclear-energy industry

*Waste management* is the general term given to the procedures and policies aimed at limiting the amount of waste produced, and then at treating it, if necessary, and disposing of it. The radioactivity in waste cannot be artificially decreased, but it does decrease with time because of natural radioactive decay. There are only two basic ways of dealing with radio-active waste: if the activity is low enough, it can be released to the environment, and if not it can be stored until it has decreased to the point where it can be released, or at least to the point where control over the store can be relaxed. Combinations of these alternatives are also possible. For example, a low or moderate-activity liquid waste can be treated to give a partially purified waste suitable for immediate discharge to the environment, and a sludge which can be stored for decay and eventual release. The treatment of a radioactive waste thus converts one waste into two others which are easier to deal with by one of the two basic methods. Some wastes may be released into the environment, e.g. by burial, in such a way that the radioactive material is not immediately able to disperse. Some decrease in activity then takes place before it becomes widely spread and thus able to affect man.

*Government policy in the United Kingdom*
The United Kingdom policy for dealing with radioactive waste was developed in the late 1940s and the 1950s and formally adopted and published as a White Paper, *The Control of Radioactive Wastes*, in 1959. The objectives of the policy can be set out in three parts:

1. To ensure, irrespective of cost, that no member of the public shall be exposed to a radiation dose exceeding the dose limits recommended by the International Commission on Radiological Protection.
2. To ensure, irrespective of cost, that the whole population of the country shall not receive from the disposal of radioactive waste an average dose of more than 1 rem per person in 30 years; this is one-fifth of the genetic dose limit recommended by ICRP* as applicable to all sources of radiation except natural background and doses to patients from medical procedures.
3. To do what is reasonably practicable, having regard to cost, convenience and the national importance of the subject, to reduce doses far below these levels.

   This policy does not mean that no waste can be released to the environment, and that it must all be stored indefinitely. Indeed, the White Paper specifically recognized that all processes give rise to some waste, and that an absolute ban on releases to the environment would have the effect of totally prohibiting the use of radioactive substances in, for example, research and medicine. It would also prevent the development of nuclear

---

* International Commission on Radiological Protection.

power. The policy aims at getting a proper balance between the two basic methods of disposal—release to the environment and storage.

The policy is applied to the control of the release of radioactivity to the environment on a case-by-case basis, and each case is dealt with in two stages. The first stage is a scientific review of the available data about the wastes and the sector of the environment to which they are to be released. This review has the aim of establishing the relationship between the proposed rate of release and the consequent radiation dose to man. Environmental problems are complex and sometimes formidable, but in the great majority of cases they can be made tractable by identifying the most important pathways through the environment, and consequently the most highly exposed group of people. These pathways and groups are then called 'critical', although 'limiting' might have been a better term. If the rate of discharge were high enough, it would be the critical group exposed via the critical pathway that would reach the absolute limit of the United Kingdom policy. This limit is almost always that applying to individuals, but if the release is to a river used as a source of drinking water for a very large part of the population, then the limit may have to be the genetic one.

The available information is usually sufficient to identify the critical pathway, and often sufficient to provide an adequate quantitative link between the rate of discharge and the dose to members of the critical group. This does not imply that all environmental data are necessarily available, but rather that many of the gaps can be closed by making obviously safe assumptions. The resulting relationship between release rate and dose will then not be particularly accurate, but it will certainly err on the safe side. If this element of deliberate caution is not excessively limiting, i.e. does not cause unreasonably high economic penalties on the discharger, there is little point in refining the assessments by costly environmental studies.

For many of the proposed releases from the nuclear industry, however, it has been desirable to carry out some environmental work to confirm the critical pathways, to identify the critical groups, and to quantify some of the characteristics of both of them. The reason for these studies has not necessarily been that the doses were expected to be significant—indeed, most have been small—but rather the need to be able to assess unequivocally the levels of dose resulting from the disposals of waste from an industry with a very high potential for releasing radioactivity to the environment.

The second stage in the application of the government policy is an administrative one. All release of radioactive waste to the environment, other than in defined trivial amounts, is prohibited unless it has been the subject of written authorization issued in England by the Department of the Environment and, in the case of nuclear sites including nuclear research centres, the Ministry of Agriculture, Fisheries and Food. Similar arrangements apply in other parts of the United Kingdom. In issuing these authorizations, the departments take account of the scientific appraisals, and thus ensure compliance with the limits laid down in the White Paper, but in addition they also consider how difficult and expensive it would be for the disposer to achieve lower releases than those requested. Only when the departments are satisfied that further reductions would be unreasonable do they issue an authorization. In this way they apply the third part of the government policy to keep doses well below the limits, having regard to costs, etc.

Many of the authorizations contain conditions requiring programmes of monitoring of the discharges and of the environment. One aim of such programmes is to provide data for enforcement purposes, and to this end the departments make appropriate check measurements, sometimes involving extensive environmental monitoring programmes of their own. A further aim is to test the environmental assessment made before the discharges started, and to provide data for future releases into similar environments. Past assessments can then be refined, and future ones improved. In this way, advantage can be taken of the flexible nature of the authorization procedure, and a preliminary limited authorization can be issued to cover the early period of commissioning and operation of a plant. The resulting environmental data can then be used as the basis for a less-restrictive authorization to meet the plant's needs as it comes into full-scale operation.

*Wastes from mining and ore concentration*
The first stage in the nuclear-fuel cycle is the mining of ore and the extraction from it of the required uranium. Mining itself produces little radioactive waste, but the subsequent process of ore concentration involves crushing and chemical extraction of the uranium. The radioactive daughters of uranium, including radium, are left in the crushed ore, which forms a large-volume low-concentration waste usually called *uranium mill tailings*. These tailings have a consistency of fine sand, and as the mines and mills often occur in arid areas, there is sometimes wind erosion. There are also liquid wastes from the extraction process, and these are usually run into settling ponds where the radioactive components settle down as a low-activity solid.

Because most uranium ores are of low quality, a large mass of ore has to be mined and processed to produce 1 tonne of uranium. The concentration plants or mills are therefore usually fairly close to the mines; as there are no uranium mines in the United Kingdom, the wastes from mining and milling do not occur in this country.

*Wastes from uranium enrichment and fuel fabrication*
Once the radium has been separated, uranium is not a particularly difficult material to handle. It is only slightly radioactive; $1\frac{1}{2}$ tonnes of uranium metal are required to make up 1 curie of alpha activity. In soluble compounds, it is chemically toxic, with a hazard rather similar to that of lead.

The process for enriching uranium in the fissile isotope, uranium-235, involves converting the uranium to a gaseous form, uranium hexafluoride. Most of the enrichment plants then use a process of gaseous diffusion in which the uranium is allowed to diffuse through a permeable membrane which allows the lighter uranium-235 atoms to diffuse more effectively than the heavier uranium-238 ones. A very small measure of enrichment is gained every time the uranium hexafluoride is passed through such a membrane. The whole plant thus consists of large numbers of pumps and diffusion barriers. More recently, attention has been given to a centrifuge process in which the different mass fractions of the uranium are separated by spinning the gaseous uranium hexafluoride at very high speeds in cylindrical centrifuge chambers. Again, the slight difference in mass allows a very small enrichment to be achieved at each stage, and the plant consists of many such stages.

The wastes from this kind of process are small, because the whole operation is conducted in a totally enclosed system. Nevertheless, plant has to be maintained, and there are small releases of gaseous waste. These are reduced by passing the air from the process areas through conventional gas-cleaning scrubbers before discharging it to the atmosphere. Small amounts of solid waste also arise in the form of worn-out plant components, although most of the uranium can be removed from these by decontamination. In many instances the metal component can then be recovered as scrap. However, the decontamination process produces small amounts of liquid waste which are released, after treatment if necessary, to local waterways.

The enriched uranium is blended with depleted uranium, and the mixture is finally converted to uranium metal or uranium oxide, depending on the type of reactor fuel that is to be fabricated. In Britain, the first generation of gas-cooled reactors use uranium metal, while the more advanced ones and the water-cooled reactors use oxide.

The chemical and fabrication stages produce small amounts of activity in gaseous waste and significant amounts of both liquid and solid wastes. The liquid wastes from the British plant are discharged to a tidal estuary where there is sufficient dilution to ensure that the small amounts of radioactivity remain harmless. There is some reconcentration of the uranium on to mud banks, but the concentration of activity in the mud is not high enough to cause a problem. The solid wastes are buried in a worked-out clay quarry; when this is full, it will be covered over and landscaped. The concentration in the buried waste is no higher than that common in ores, so there are no special environmental problems. As a precaution, however, a check is kept on uranium in the drainage from the quarry, although this does not lead to drinking-water sources.

*Wastes from reactors*
The wastes from nuclear reactors are, generally speaking, rather minor and their characteristics depend on the type of reactor. For the gas-cooled reactors that form the basis of the British nuclear-power programme, there are small releases of radioactive gases to atmosphere, partly due to the activation of the noble gas, argon, which occurs naturally in air, and as an impurity in the carbon dioxide coolant of the reactor. Air is used to remove heat from parts of the shielding of the earlier designs of reactor, and there is always some slight leakage of carbon dioxide from the coolant circuit. All the reactors therefore release a small amount of the argon isotope, argon-41. Two other noble gases, xenon and krypton, are amongst the fission products; since no fuel is entirely without defect, traces of these gases get into the coolant, and a small fraction is then released to the atmosphere. With one exception, krypton-85, these noble gases have short half-lives and do not accumulate in the atmosphere. They do not play any part in environmental or metabolic processes, but they do cause small radiation exposures of man, merely by being present in the surrounding air. The level of this dose to people in the area immediately round a reactor varies from a few per cent of the dose from natural background down to a fraction of 1%, depending on the type of reactor.

When the uranium fuel has been irradiated in the reactor to the point where its nuclear and mechanical properties have deteriorated sufficiently, it is discharged from the reactor to await reprocessing. The first stage of this period takes place in water-filled storage ponds at the reactor site. To save storage space, some of the purely mechanical parts of the fuel elements are removed; since these have been exposed to neutrons in the reactor, they have become activated, and have to be treated as solid wastes. They are stored in concrete-lined vaults on the reactor site, where they can remain until the relatively short-lived activation products have decayed to trivial levels. The fuel elements in the storage ponds inevitably suffer some slight corrosion, although care is taken to keep this to a minimum. The pond water then becomes slightly active, partly as a result of activation products in the fuel cladding, and partly as a result of fission products. These arise from traces of uranium on the outside of the fuel elements, and from uranium exposed within the elements as a result of occasional imperfections in the cladding. The water in the ponds is recycled and purified, but eventually some of it has to be discharged to maintain the clarity, and to prevent activity building up in the water to the extent where it would become an operational problem on the plant. The treated pond water is therefore mixed with the large volume of coolant water required by the power station's condensers and discharged to the environment. In all but one case in Britain the discharges are to coastal waters. The one exception is the nuclear power station at Trawsfynydd in North Wales, where the discharges are made to a freshwater lake.

Detailed appraisals have been made by the Ministry of Agriculture, Fisheries and Food of the significance of these discharges, and they are subject to extensive monitoring programmes. The results of these programmes are published routinely by the Ministry and show that the radiation doses to even the most highly exposed people are less than 1% of the dose limit recommended by ICRP, i.e. only a few per cent of the natural background. Because of the more limited dispersion available in the lake, and also because of a concentration process of caesium in freshwater fish, the doses to a few people in the area round Trawsfynydd are somewhat higher, amounting to a few per cent of the ICRP recommended dose limit.

*Wastes from fuel reprocessing*

When the radioactivity in the fuel at the reactor station has decreased sufficiently, the fuel is transported by rail to the reprocessing plant. It contains almost all the radioactivity involved in the nuclear fuel cycle, and it is therefore at the reprocessing plant that most of the waste disposal problems arise.

The basic form of a chemical reprocessing plant is shown in figure 5.2. The incoming fuel is stored for a further period to reduce its activity; it is then separated from its cladding and dissolved in acid. For some fuels, these processes are combined. The acid solution containing uranium, plutonium and the fission products is then repeatedly mixed with and separated from an organic solvent in the process known as *solvent extraction*. In the first stage of this process the heavy metals, uranium and plutonium, are transferred to the solvent phase, and the aqueous solution then contains almost all the fission products. The chemical form of the plutonium in the solvent phase is modified, and it can then be extracted into an aqueous phase in a further stage of solvent extraction, leaving the uranium in the solvent. Finally, the uranium can be extracted into an aqueous phase in a third stage of separation. Both the uranium and the plutonium still contain traces of fission products, and these are removed by successive purification stages. The purified uranium is low in the fissile isotope, uranium-235, but it can still be blended with enriched uranium, or itself re-enriched, and then returned to the fuel cycle. Initially, the plutonium is stored in solution, and subsequently it is converted to plutonium oxide or to a mixed uranium-plutonium oxide for use as a fuel in fast reactors. For this purpose the oxide has to be formed into pellets, which are then encased in stainless-steel cladding and built up into fuel

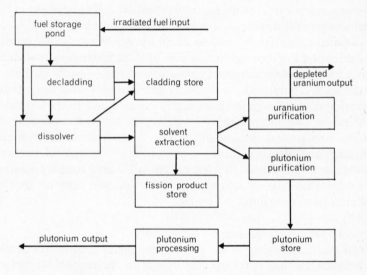

**Figure 5.2**   The irradiated-fuel reprocessing plant.

elements for use in the reactors. At present, the only plutonium-fuelled fast reactor in Britain is the Prototype Fast Reactor (PFR) at Dounreay in the north of Scotland. A small amount of specially prepared plutonium is converted to metal for use in the British nuclear-weapons programme.

*The Fission Product Wastes.* The aqueous fission product stream from the first cycle of solvent extraction is extremely radioactive and constitutes one of the major wastes of the nuclear-energy programme. Although highly active, it is still chemically very dilute, and it is therefore concentrated by evaporation before being sent to storage. The storage tanks are double-walled stainless-steel tanks inside concrete enclosures. Spare tanks have to be provided in case a leak develops, though none has occurred yet. The space occupied by these tanks is surprisingly small, and they will occupy less than 2 acres (less than 1 hectare) even at the end of the century.

One problem posed by storage of fission products is the amount of heat released. The process of radioactive decay releases small amounts of heat, typically about 5 milliwatts/curie of fission products, but the number of curies in the fission product store is so large that the heat yield is very considerable. A newly-filled tank may require the removal of as much as 2 megawatts of heat. This is achieved by building cooling coils into the tank and by passing cold water through them. The intense radioactivity also causes some dissociation of the water into hydrogen and oxygen, and this requires that a continuous flow of air be passed over the surface of the liquid in the tanks to ensure that there is no build-up of an explosive mixture of these gases.

The fission-product storage tanks then require continuous attention, and provision has also to be made for pumping the liquid out of the tank into the spare should a leak develop. These processes are not difficult, but they do require continuous maintenance of a chemical-engineering character. It is therefore likely to be preferable to convert the liquid to a more stable form of material for storage, and the most promising seems to be a form of glass. Enough work has been done at large-scale pilot-plant level to show that such a process would be feasible and reasonably economic. The glass material containing the fission products would be formed inside stainless-steel containers, which would themselves then have to be cooled, probably by storing them under water in an artificial pond. Watch-keeping and some minor maintenance would be required, but this would be much simpler than for the liquid store.

Whatever storage process is adopted, it will need to be continued for a very long time. The fission products themselves need to be stored for several hundred years before the liquid or glass can be reasonably disposed of into the environment, either by burial in deep geological formations or by placement on the bed of the deep ocean. By that time the fission-product activities would be low enough for either of these two solutions to be perfectly safe. Unfortunately, the fission-product wastes also contain the elements of higher atomic number than plutonium produced by multiple processes in the reactor. Some of these materials have very long half-lives, and some decay back to plutonium isotopes, which are also very long-lived. They therefore remain in the fission-product wastes even when the fission products have themselves decayed, and they severely complicate the ultimate disposal of this waste. Since they do not reach a negligible activity for periods of the order of 100,000 years, storing them is really no solution and is merely a way of buying time in the hope that future generations will have satisfactory permanent solutions to offer. This is sometimes regarded as an unacceptable weakness of the present nuclear-energy programme based on fission, and it certainly poses a very real moral problem. It is perhaps not as bad as it is sometimes painted, because we are not transferring to future generations a problem for which there is no adequate interim solution. What they have to do is to continue to store the waste in the same way that we have been doing. A failure to continue the storage under proper conditions will certainly result in a local hazard, but if the waste is in a satisfactory form, such as glass, it will not be a disaster. If civilization has collapsed to the point where it cannot maintain control over these wastes, then their existence will not add significantly to the problems then facing mankind.

However, this 'wait and see' attitude is not in itself sufficient, and a considerable amount of work is now being done on the possible ultimate disposal of these long-lived alpha-active wastes. It may be possible to separate them from the fission products and to convert them to very much shorter-lived materials by irradiation in reactors. This would require major technological developments and may not be feasible. Alternatives under investigation include the possibility of disposal of the material, perhaps in a glass matrix, on to the bed of the deep ocean. The activity would remain in the glass for very long periods and, when it did eventually begin to go into solution, there would be a very large volume of water available for dilution. The chemistry of these materials also suggests that they would be rapidly attached to particulate material and deposited on the sea bed. We already know enough about the physical and chemical properties of the oceans to be sure that substantial amounts of radioactivity can be disposed of in this way with complete safety, but there are obvious problems of political and public acceptance, and a need to take great care to avoid compromising future developments of exploitation of sea-bed minerals. However, the volumes of waste are small and the ocean floor large, so that this last problem should not prove too serious.

The alternative approach to natural isolation is that of deep and stable geological formations. The glass matrix would keep the radioactive materials separate from the environment for very long periods, and it seems likely that geological formations can be identified which will not be subject to major mechanical movement in periods approaching 1 million years. Most of these long-lived materials appear to be very immobile in soils and geological materials, and their transfer back to man after they had been released from the glass would be exceedingly slow, if indeed it occurred at all.

In short, while it is certainly true that the problems of very long-lived alpha-active wastes have not yet been fully solved, it is probably an exaggeration to describe our continued production of these wastes as irresponsible. Storage is a perfectly satisfactory method of dealing with them, provided that there is some continuing co-operation from future generations. Of the three ultimate solutions under investigation, the geological solution seems the most likely to be both feasible and acceptable, although deep-sea disposal would probably be cheaper and, in the long run, safer if the political problems could be overcome. The process of 'burning' the material in a reactor is by far the most satisfactory, because it genuinely destroys the radioactivity in the waste, but the technological difficulties seem to be substantial.

*Gaseous Wastes.* Probably the two most important gaseous wastes from a fuel reprocessing plant are the gaseous fission products released when the fuel is dissolved, and the entrained fission products associated with the evaporation and storage of the fission-product liquid wastes. Because of the prolonged cooling period between irradiation in the reactor and reprocessing—typically about 6 months—the short-lived gaseous and volatile fission products have largely decayed before the fuel is dissolved. The important exception is krypton-85 with a half-life of 10·76 years. This noble gas is released from the fuel at the time of dissolving and discharged to the atmosphere. It is then dispersed by meteorological processes, and eventually becomes mixed fairly uniformly with the whole atmosphere, though remaining predominantly in either the northern or the southern hemisphere depending on where it was released.

Krypton-85 is predominantly a beta-emitter, though it also emits a gamma-ray in about 1% of its disintegrations. The dose it delivers is therefore predominantly to skin, although the gamma-ray does cause some irradiation of the whole body. Because of its long half-life and wide dispersal, the krypton in the atmosphere is a mixture of that produced by all the reprocessing plants; the resultant exposure is fairly uniform across the whole northern hemisphere. There is an additional contribution from locally-released krypton to those people who live near a large reprocessing plant. Because of this very widespread distribution, and of the very large number of people who are exposed as a result, it has received a good deal of attention. The radiation doses are much too small to measure directly, but they can be calculated from a knowledge of the releases and of the meteorological processes by which the krypton is dispersed. Even by the end of the century, when the world nuclear-power programme will amount to perhaps 3 million megawatts, the gamma-ray dose from krypton will be less than $\frac{1}{1000}$ of the natural background dose.

The skin dose from beta-radiation to people in the immediate vicinity of a very large chemical reprocessing plant would be higher, but would still be substantially less than natural background unless the plant were in an area of unfavourable local meteorology. Despite these low doses, there has been a good deal of concern about the releases of krypton, and development has been carried out of ways of removing it from the gases leaving the fuel dissolvers. In addition to air, these gases contain acid fumes and oxides of nitrogen, as well as fairly large volumes of stable xenon which is produced in the fission process. A multi-stage process would certainly be needed, probably with a final low-temperature stage to separate krypton from xenon, and thus simplify the storage problems. The krypton has a short enough half-life for it to be stored for some decades and then released, thus avoiding the indefinite build-up of storage cylinders or tanks.

By contrast, the entrained fission products present a much simpler and more localized problem. They are to a large extent removed from the gas streams by conventional gas-cleaning methods, such as scrubbing and high-efficiency filtration, and the traces that remain are then discharged from high stacks. By the time they reach ground level they are extremely dilute and give rise to almost undetectably small doses. One of the ways in which they might return most effectively to man would be by deposition on grazing land and transfer by the cow into milk. The routine monitoring of milk therefore forms an essential part of the control programme round a separation plant, although the results are consistently negative in normal operations.

*Liquid Wastes.* Liquid wastes arise at a reprocessing plant from most stages of the operation. Large volumes arise from the storage ponds where the fuel elements are received from the reactors, but the amount of radio-activity in these wastes is small. The solvent extraction process used to separate the uranium, plutonium and fission products, also gives rise to liquid wastes, because the separations are never perfect. In particular, traces of fission products have to be removed from the uranium and plutonium in special purification stages. The more active of these wastes, sometimes called intermediate-activity wastes, are concentrated by evaporation, and the concentrates added to the main high-activity fission-product wastes. The condensates are of low activity, but are not completely pure, and these form a large-volume low-activity waste. Altogether well over 99·9% of the fission products find their way into the main fission-product store, but a fraction of 0·1% appears as large-volume liquid wastes.

At the British separation plant on the Cumbrian coast, these wastes are discharged by pipeline into the Irish Sea. The process was the subject of very detailed studies in the late 1940s up to the time when the plant first became active, soon after 1950. In the early stages of operation, only small amounts of activity were released to the sea, and very detailed studies were carried out to relate the amount of discharge to the level of radioactivity in various marine materials, including fish and seaweed. As the results of the preliminary studies were confirmed and refined by the data coming from these preliminary discharges, authorization was given to the plant to increase the amounts. Over the subsequent 20 years or so, the throughput of fission products has increased by a factor of around 30, but the releases of activity to the sea have increased by a factor of only about 3. This has been achieved by increasing the effectiveness of waste management on the plant, so that a progressively higher proportion of the fission products goes to the highly-active fission-product store, or is delayed on the plant for partial decay prior to discharge.

There are two critical pathways, and hence critical groups, associated with these discharges. The first is a reconcentration of fission products, particularly ruthenium-106, on to a form of edible seaweed, *Porphyra umbilicalis*, that grows along the Cumbrian coast. This weed is harvested and shipped to South Wales, where it is processed and sold as a delicacy known as laverbread. Many people eat small amounts of laverbread, and a small number quite considerable amounts. This small number constitutes the critical group. Over the years their exposure has amounted to about 10% of the dose limit recommended by the International Commission on Radiological Protection.

The second critical pathway has been the deposition of activity on silt that builds up in a small local estuary near the discharge point. One or two people who work over the silt beds collecting bait for fishing are exposed to gamma-radiation from the silt, and their level of exposure also may be as much as 10% of the ICRP dose limit.

As might be expected, the doses to these critical groups are somewhat larger than those occurring around individual power stations, but they are not increasing in proportion to the rate of increase of nuclear power and, as the nuclear-power programme extends still further, additional improvements in waste management at the reprocessing plant will be called for, and further increases in the releases of radioactive waste to sea seem improbable.

*Solid Wastes.* All the stages of the fuel cycle produce solid wastes, but those from the reprocessing plant are perhaps the most important. It is convenient to think of them in three separate classes, although there is inevitably some degree of overlap.

(1) The most important solid waste is undoubtedly that made up of the fuel cladding which has been separated from the uranium fuel. This cladding is highly active, both as a result of activation by neutron irradiation and because of plutonium and fission-product contamination from the fuel. The cladding material is stored in concrete blockhouses, known as *silos*, within the confines of the separation plant. The activation products,

and most of the fission products, have reasonably short half-lives, so that almost all the activity will decay in the lifetime of the silo. This is not true, however, of the longest-lived fission products nor of the plutonium, and ultimately the material will have to be recovered from the silo for further disposal.

(2) The second type of solid waste is that contaminated by fission products, and consisting of miscellaneous trash from the operation and maintenance of the chemical plants. It contains very little plutonium and is transferred to a nearby site, where it is buried in large deep trenches. Control will be maintained over this site until the fission-product activity has decayed to the point where control can be relaxed safely, and the site landscaped.

(3) The third class of material is a similar kind of trash coming from the plutonium plants. This contains too much plutonium to allow it to be buried, since it would outlive any reasonable period of control over the site. It poses the same problems as the long-lived alpha-active material in the fission-product wastes, but, because it is already free of fission products, it is available for immediate processing or disposal if methods are available. Some of the plutonium in this trash might be recovered, although with a material of such variable composition any recovery programme would be difficult and expensive. Inevitably, even after recovery, some plutonium will remain in the waste. The disposal solutions available are exactly the same as those already discussed for the fission-product wastes. For this material and for the fuel hulls, once the activation products and fission products have decayed, no final choices have yet been made.

### The consequences of nuclear accidents

Very great care is taken in the design, construction and operation of nuclear reactors to make accidents unlikely. Nevertheless, human perfection does not exist, and there remains the possibility, amounting in the long run to a certainty, that various kinds and severities of accident will occur.

Common sense and experience to date indicate that most accidents are small, but that a few may be serious. From the point of view of the environment, most accidents to nuclear reactors are indeed of no consequence at all, although they may well be of considerable economic importance to the plant. Both in Britain and America, a great deal of effort has been applied to assessing the various sequences of events that might lead to a release of radioactive material from a reactor, and thence, through the series of barriers designed and built into the reactor structure, into the environment. These studies have contributed to reactor safety by identifying the most important design and construction features.

The essence of reactor safety is care and attention to detail at all stages, starting with design and proceeding to construction (in which quality control plays a major part), inspection, and finally to operation and maintenance, which themselves are also subject to inspection. There is a legitimate fear that operator error or malice might be the prime cause of a serious accident, but this particular risk is one that can be reduced almost indefinitely by good design.

Nevertheless, after all reasonable, and some unreasonable, precautions have been taken, there is still a small but finite risk of an accident serious enough to release fission products to the environment.

Essentially, any such accident involves overheating of the fuel. This may be initiated by a loss of the reactor coolant as a result of mechanical failure of the coolant circuit. The nuclear reaction will be shut off by one of several emergency mechanisms, but the residual heating effect of the radioactive fission products may still be enough to cause a failure of part of the nuclear core. The gaseous fission products and part of the volatile ones, such as iodine and caesium, will be released. Depending on the reactor design, some part of these, often the major part, will be retained by secondary containment systems, but these will not be perfect, and some release to the environment must be expected.

Every reactor installation is required to have a detailed emergency plan. This will be adequate to protect individuals in the environment against any likely accident. In this context, 'likely' means having a probability of occurring more than about once in a period of between 1000 and 10,000 years of operation of the reactor. The public might be inconvenienced by being asked to leave their homes for a few hours, or even days, and some food products, especially milk, might have to be destroyed, but no one would be at any real risk.

There remains, however, the faint possibility that much more of the core might be damaged, and that the secondary containment devices might fail at the same time. The emergency plan would then be overwhelmed, and severe radiation exposures would occur at distances up to perhaps a few tens of miles. Some radiation casualties might occur, and over the next 20–40 years there would probably be some increase in premature deaths due to cancer. It is unlikely that these deaths would be statistically detectable, but they would be nonetheless real.

Such an event would be extremely serious and must therefore be made extremely rare. Current estimates suggest that it would have a probability of no more than one in 1 million years of operation of a reactor. The chance of being killed by such an event, even for people living close to the reactor, is then extremely small, and is no more than the risk of being killed by a meteorite. It is about 100 times less than that of spending 1 day a year as a pedestrian in a typical city.

Perhaps even more serious than the actual casualties would be the disruption of normal life. An area of some tens or even hundreds of square miles might need to be decontaminated before it became fit for prolonged habitation, and there would be substantial effects on agriculture over even larger areas.

Such an accident would be exceedingly damaging and exceedingly expensive, but it would fall short of the scale of many natural disasters. Its toll would probably be less than that caused by an aircraft crashing on an occupied sports stadium. If the risk of its occurrence can really be kept down to about 1 in $10^6$ years of reactor operation, such an accident would occur on average about once in 3000 years on the present scale of nuclear energy, rising to about once in 100 years by the end of the century. Even on a world-wide basis this is not negligible, but it is difficult to describe it as intolerable. The emphasis in reactor safety must clearly be, and continue to be, on the quality of design, engineering, operation and inspection.

### Centres of apprehension

Nuclear power has been a subject of controversy in varying degrees all over the world, but the form and intensity of the controversy has been remarkably variable. Britain, which moved ahead earlier and more rapidly than other countries, and thus has the highest proportion of nuclear power in the world, has had remarkably little public alarm, while the United States has had the most dramatic confrontations. Because of the intensity of the American worries and because they have covered all the topics of interest, it is convenient to treat them as characteristic of the nuclear apprehensions all over the world.

The controversy about nuclear energy has been partly scientific, and partly economic and political. There have been worries about the consequences to health of the routine effluents from nuclear power stations, and of the much larger releases from possible major accidents. There has been a basic objection to nuclear energy, merely because it is an additional source of energy and thus makes possible continued economic growth, which itself is anathema to some economists. There are worries about the ultimate disposal of the long-lived radioactive wastes, and about the problems posed by nuclear power stations when they have come to the end of their useful life. Finally, there is a real fear that the materials of nuclear energy, especially the plutonium, make a world nuclear war more likely, and put disastrous powers within the reach of terrorist organizations.

It is not always possible to identify the real worries in a public controversy, because use will be made of any effective weapon. Nor is it easy to tell whether the reaction is genuinely widespread, or confined to a small but vocal group of enthusiasts. In these respects, the nuclear controversy is no different from any other major area of public contention.

### Releases of waste to the environment

These were an early source of alarm, and pressure was applied in the United States to make the radiation standards more restrictive. One argument was based on the contention that one particular figure in the current standards would permit a situation that would kill some 30,000 Americans every year. In fact, the application of other parts of the standards ensured a figure at least 1000 times smaller, and a more widely-held view of the biological consequences of low levels of exposure suggested an ultimate casualty rate of between 0 and 3 deaths per year in the US population of over 200 million. In the event, the US Atomic Energy Commission clarified the application of its standards without changing the basic figures.

The fact that public exposure from reactor effluents is below, and will remain below, the exposure from natural background radiation has decreased the interest in these effluents and concentrated attention on the much larger doses from possible accidents.

### Reactor accidents

The controversy about accidents has been concentrated in two areas— the adequacy, or inadequacy, of the quality of the design and engineering of reactors, and the consequences of very severe accidents. In the United States, the initial approach was to define a 'maximum credible accident' and to provide engineered safeguards to ensure that such an accident would have only small consequences. This simple approach has given way to a policy developed in Britain, in which all possible accidents are considered and studies made to establish their likelihood. The acceptability of a reactor then depends on a combination of the severity and the improbability of its potential accidents.

This more sophisticated approach is apparently finding more support among informed journalists than did the earlier one. It is not easy to assess the public response—possibly because the public finds it difficult to take seriously any event as unlikely as being hit by a crashing aircraft or a meteorite.

*Long-term waste storage*

This is one of the most real, but perhaps the least widely felt, worries about nuclear energy. There is a genuine ethical dilemma in passing on to future generations problems for which we now have only interim solutions. We can tell them what to do to control the problems, but we commit them to continue to follow our advice indefinitely unless they can devise a permanent solution. It is, of course, not our only ethical dilemma concerning future generations. They may well look back on the three or four generations which used all the world's oil as a fuel rather than as a chemical resource with some justifiable anger.

*Reactor and plant decommissioning*

The nuclear-energy industry is so new that none of its major plants has yet had to be taken out of service and got rid of. Some experimental and demonstration reactors have been closed down, and a few have been fully dismantled, but none of the big plants has had to be dealt with in this way. Despite some worries that such a move would prove impossible, the real question seems to be economic. The methods of dismantling the plant, including the highly active structure of a reactor, are already available. They would be expensive if they had to be applied at once, as soon as the fuel had been discharged, and would get progressively easier and cheaper over the years as decay reduced the residual activity. The appropriate time for dismantling would thus depend on the economic and amenity value of the site. There would be a substantial production of solid wastes, but these would have characteristics similar to those arising from other parts of the nuclear fuel cycle and could be disposed of appropriately.

*Nuclear warfare and terrorism*

A country with a nuclear-energy programme can eventually become a country with nuclear weapons. The transition is not easy, and there is a massive system of international inspection to make it less likely, but it remains possible. The controversial point is then: do we avoid using nuclear energy and thus diminish the risk of nuclear war? The arguments cover an enormous range of subjects. Perhaps the availability of energy decreases world tension, and thus reduces the chance of a nuclear war. Perhaps the highly-developed countries have no moral right to attempt to withhold a new energy source from the less-favoured nations. Perhaps an industrial technique once demonstrated cannot be suppressed. At all events, unanimity does not seem to be within reach on this subject.

The diversion of nuclear materials by terrorists has been a popular subject, both for discussion and for fiction. The creation of a serious public hazard using material hijacked in transit would be a fairly difficult operation, and the construction of even an inefficient nuclear weapon very much more difficult than many accounts have implied. Logic suggests that there are more effective ways for terrorists to operate, but logic is not necessarily the right criterion. Meanwhile, the nuclear industry continues to take precautions and, one hopes, to keep secret the more important aspects of these precautions.

### The final judgment

We are much too close to the situation to be able to make any final judgment about the long-term implications of our present decisions on nuclear energy. However, a few questions can already be answered with reasonable confidence.

*Are we and our environment at risk now?*
From the normal operations of nuclear power stations—no, unequivocally no.

From possible reactor accidents—yes, to a small extent. The disastrous accident now seems to be very very unlikely; the smaller, still unlikely, accident would be a nuisance rather than a real danger.

*Shall we be at greater risk if we continue to use nuclear energy for decades?*
Obviously the accident risk must increase with increasing numbers of reactors, unless technology develops compensating improvements. The routine operation of reactors remains a very small problem, but some increased application of existing technology will be needed to keep the effluents from reprocessing plants at or below their present levels.

*Do we have a tiger by the tail?*
Yes, in one sense, we do. Embarking on any highly-developed industrial civilization is an almost irreversible decision. We do not know how to revert to a low-technology agrarian community. We do not know if we want to try. In this sense, our whole life is aimed at keeping up with our tiger—civilization. Nuclear energy is just one part of that tiger.

More specifically, however, we have to consider whether long-lived nuclear wastes cause a situation which is so objectionable that we are not morally justified in creating it. For some people, that tiger is very real; for others, it is already tamed and its longevity is no more than a minor embarrassment. Here, especially, the future will have to judge.

*Do we need an energy policy?*
Yes, indeed we do. No source of energy is free from disadvantages. No choice of one fuel rather than another is straightforward. The factors are

economic, usually and perhaps wrongly given first priority, social, political, environmental, conservationist and moral. We must not consider one energy source or one factor in its choice in isolation. Indeed, we should not consider energy itself as isolated from broader issues, such as population, food supplies, world resources and political stability.

Without doubt, establishing an energy policy is an enormously difficult task. But tasks like this are, surely, what modern government is all about.

## FURTHER READING

*Atomic Energy Waste* (Glueckauf, E., ed.), Butterworth, London, 1961. A comprehensive review of the problems, with many technical data for the specialist.

*Environmental Aspects of Nuclear Power Stations*, IAEA, Vienna, 1971. A collection of symposium papers and discussion, giving the general reader a broad background. To be read selectively.

*Environmental Surveillance around Nuclear Installations*, IAEA, Vienna, 1974. A collection of symposia papers of general interest.

*Management of Radioactive Wastes from Fuel Reprocessing*, OECD, Paris, 1972. A collection of symposium papers, generally very readable, and containing many technical data.

National Radiological Protection Board, *Living with Radiation*, HMSO, London, 1973. A short account of radiation and its implications for the general reader.

Seaborg, G. T. and Corliss, W. R., *Man and the Atom*, Dutton, New York, 1971. A popular, but not impartial, account of the potentialities of nuclear energy.

*Waste Management Practices in Western Europe*, OECD, Paris, 1972. A short but thorough review for the general reader.

CHAPTER SIX

# DIRECT USE OF SOLAR ENERGY

B. J. BRINKWORTH

## Introduction

In the autumn of 1973, the world experienced a shortage of oil, followed
by a sharp increase in its price. These events caused many people to give
serious consideration for the first time to the need for conservation of
energy, the advocates of which had, in the prior era of cheap fuel, been
largely unheeded. In later years, this time may well be viewed as a critical
point in history, and the experiences that went with it as a blessing in
disguise; for, although this oil crisis was politically inspired, and was not
due to demand outstripping the supply, it foreshadowed a real shortage
which, though not far off, was not being anticipated seriously. As a result,
there now exists an atmosphere in which it is possible to discuss the
problem of long-term energy supply with some prospect of action being
taken.

Predictions of the probable reserves of fossil fuels and the periods for
which they will last have always been difficult. However, it seems now to
be generally accepted that the peak of oil production may occur within
20 years or so. Coal will be available for much longer, and the way that
it is used will have to change, so that it takes on some of the functions of
oil. These include the furnishing of organic chemicals and even of liquid
fuels, which will always be needed for certain vehicles, notably aircraft.
Nevertheless, the end of the era of the fossil fuels is clearly foreseeable,
and we may need to take action now to ensure a continuation of energy
supply as they begin to be worked out. This is a formidable problem.
Already the energy demand in the more developed nations has reached
some 70,000 megajoules (MJ) per year per capita. To bring the whole world

up to this level would increase the total demand by a factor of at least 5 at the present time, and by the end of the century the increase in population would raise it by a further factor of about 3.

In this situation, it is hard to be unimpressed by the timeliness of the discovery and exploitation of fission energy. Nuclear power seems to be potentially capable of averting a world energy starvation, and is being seen by many in the developed countries as the most promising technical option in this field. Yet it is not universally welcomed. The scale on which reactors would have to be built, the problems of waste disposal, thermal pollution, and of safety and security are causing disquiet. Even the possibility of an early shortage of uranium has been raised. It is not surprising, therefore, that recently much interest has been shown in energy sources which are substantially inexhaustible. We cannot live indefinitely on our capital resources, and ultimately we must learn to live within our income. Nearly all the world's energy income comes from the sun, via the solar radiation. This energy is responsible for warming the land and the seas; for driving the winds (and hence the waves) and the ocean currents; for evaporating water, which returns as rain; and for driving the photosynthetic reactions in plants which provide food for the entire animal kingdom. The ultimate fate of this vast tide of energy is to return as long-wavelength radiation back into space. Many have asked whether we could use it more directly for human purposes while it is here. The energy in the radiation which falls on a region a mere 100 km square in the tropics exceeds all that required by man at the present time; no possible human population would need more than falls on the Sahara desert. These beguiling statements reveal both the attraction and the drawback of solar energy. The areas are small compared with the surface area of the earth, but they are large in terms of human constructions. A 100-km square is comparable with the surface area of all existing buildings. Plainly, we cannot lightly contemplate the construction of collectors for this energy if it means undertaking a task equivalent to erecting all the world's buildings again. In preparing to utilize solar energy, we are thus facing a paradox: the energy is available in an ample and inexhaustible supply, but we may not be able to collect it.

It is instructive to note that we had almost reached this point a century or more ago, but were able to circumvent it by learning to use the energy stored in concentrated form in coal and oil. In the pre-industrial era, solar energy was the predominant source of power, for the windmills, watermills, and sailing ships of the time were driven by it. But these could not have powered the subsequent period of industrial expansion, because they could not have been made large enough. They were well suited to modest power requirements and local needs, but could not serve the demands of intensive industry, concentrated in the interests of high productivity. The watermill has survived, in the form of the hydraulic turbine, and a renewed interest is being shown in the development of windmills, using modern technology. Attempts to use solar energy directly, either for heating or for driving a thermodynamic system, were well in hand a century ago, but it was clear that these could never be competitive in an era of cheap fuel. As this era approaches its end, it is these direct uses which are receiving the most attention.

Here, then, is an energy source of great appeal: ample, inexhaustible, silent, non-polluting. It is important that there should be general awareness of what we might expect from it—and of what we might not.

### The raw material

Thermonuclear reactions in the core of the sun, in which the nuclei of hydrogen atoms are being fused into those of helium, are causing a net loss of mass of some four million tonnes per second. Most of the energy thus released is carried away by very-short-wavelength photons. As these pass outward, they interact with other nuclei, sharing their energy so that the re-emitted photons are less energetic and hence of longer wavelength. At the visible surface, the photosphere, the temperature is sufficiently low for the matter to be mostly in the form of whole atoms. The photons which leave this region thus have an energy distribution closely resembling that for a classical black-body radiator, with a temperature of about 6000 K. This distribution, relative to photon wavelength, is illustrated in figure 6.1, which shows that about half the energy (represented by the area under the curve) is carried by photons in the visible region, and most of the rest in the infra-red. This makes the sun an unusual radiative source, for no terrestrial sources (other than nuclear explosions) yield much energy at such short wavelengths.

From an engineering point of view, the sun is also a feeble source. When the radiation reaches the earth's orbit, about eight minutes after leaving the sun, it has spread out into a region about 150 million km in radius. The total energy flux is then about $1.35 \, \text{kW/m}^2$. Though not inconsiderable, this is thousands of times smaller than has been accommodated in recent engineering practice. Even the flux at the surface of a domestic kettle element is a hundred times as great.

There is substantial interference with the solar radiation during its passage through the atmosphere. At the higher levels, above an altitude of about 25 km, oxygen molecules are dissociated into atomic oxygen by the absorption of high-energy photons. Some oxygen atoms combine with molecules to form ozone, which also absorbs photons and is dissociated again so that at any time there is a steady state in which $O, O_2$ and $O_3$ co-exist, high-energy photons are absorbed and low-energy photons emitted. These processes remove virtually all the radiation with wavelengths less than about $0.35 \mu m$, that is, the ultra-violet. Photons of the visible and infra-red radiation do not possess enough energy to produce dissociation, but are subject to a variety of absorbing and scattering processes in the lower atmosphere. Absorption occurs mostly in molecules of water vapour and carbon dioxide, at certain characteristic wavelengths, mostly

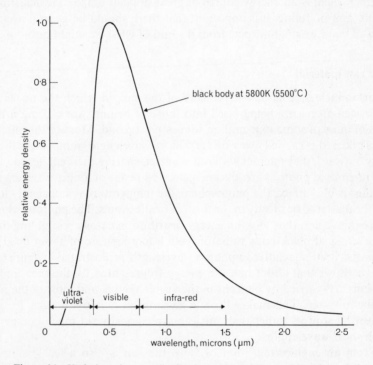

**Figure 6.1**   Variation of energy distribution with wavelength (SEM, figure 2.1).*

in the infra-red. The energy removed from the solar radiation helps to warm the atmosphere and is then re-radiated at very long wavelengths, greater than $10 \mu m$, so that some of it reaches the surface by this route.

Scattering processes do not change the wavelength of the radiation, but alter its direction of propagation. Some scattering occurs on the molecules of the air itself; the preferential scattering at short wavelengths is responsible for the apparent blue colour of the clear sky. Fine solid particles in suspension also cause scattering at predominantly short wavelengths, so

* Brinkworth, B. J., 1972, *Solar Energy for Man*, Compton Press.

that the haze due to the presence of dust and pollutants frequently lends a blue coloration to the distant view. Water particles in mists, clouds and fogs are larger than the wavelengths of light and scatter it differently. There is less wavelength-dependence, so that the scattered light seems white; and, instead of being spread fairly evenly in all directions, most of it is deflected only a little from the original direction of propagation. Thus,

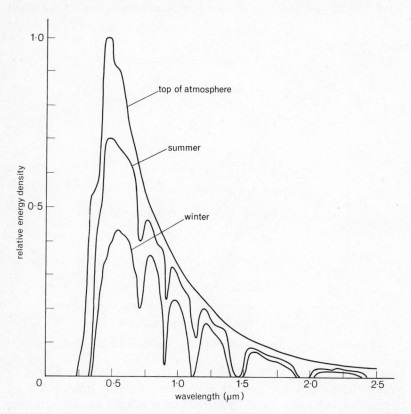

**Figure 6.2** Effect of passage through atmosphere on solar radiation (estimate for southern UK).

although scattering by the droplets in clouds is a significant means of re-directing solar energy back into space, a surprisingly large proportion is transmitted through clouds, except when they are very dense and deep.

Figure 6.2 illustrates the net effect of interference by the atmosphere with the solar radiation reaching the earth's surface. When the sun's elevation is low, and the radiation has a long path to travel through the atmosphere, the effects are substantial. The general reduction is caused by scattering processes, and the local fissures by absorption at particular wavelengths. Even when the sun is directly overhead, the energy flux at the surface rarely exceeds $1\,\text{kW/m}^2$.

It is well known that the earth's axis is inclined at an angle of about $23\frac{1}{2}°$ to the perpendicular to the plane of its orbit. As a result, the maximum altitude of the sun at a point whose latitude is $L$ is $90° - L + 23\frac{1}{2}°$ (about $60°$ in Britain) and at the time when this occurs (midsummer) the day is

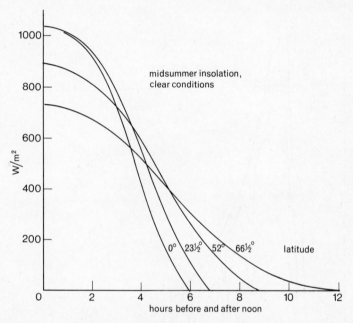

**Figure 6.3**   Energy flux on horizontal surface (insolation) in clear conditions.

longer than the night. Typical variations of the energy flux on a horizontal surface (the insolation) at various latitudes are shown in figure 6.3. It will be seen that as latitude increases and the solar altitude decreases, the day length increases. These have opposing effects on the total energy reaching the surface (proportional to the areas under the curves of figure 6.3) and this is found to be surprisingly little dependent upon latitude, right up to the polar circles. A clear day in midsummer yields about $30\,\text{MJ/m}^2$ on a horizontal surface at most latitudes. In the winter, on the other hand, the maximum solar altitude is $90° - L - 23\frac{1}{2}°$ (about $15°$ in Britain) and the day shorter than the night. As latitude increases, the solar altitude decreases and so does the day length. This combination produces a very marked effect of latitude, and the clearest day in midwinter yields only about one-tenth of the midsummer value at the latitude of Britain, and nothing beyond the polar circles. At this time of year there is a significant difference in the values between the south of England and Scotland, though this is not so noticeable in summer.

Superimposed on the fluctuations with time of day and season of the year are those due to the weather. Cloud cover, particularly near coasts and rising ground, causes random variations in the energy reaching the surface. Because of the preferential forward scattering that occurs in the less-dense clouds, the reduction in the total energy is not as great as is usually imagined, the fall in the direct component being partly balanced by a rise in the diffuse component. Britain has an unusual, if not notorious, degree of cloud cover, for example, and more than half the energy received from the sun here arrives at the surface diffusely. Yet the total received by a horizontal surface in a year is about $3500 \, MJ/m^2$ and the corresponding value for Australia, which enjoys quite the opposite reputation, is about $6500 \, MJ/m^2$, less than twice as much.

Many countries operate a radiation-monitoring service, from which the data are collated at an international centre in the USSR. In Britain, the Meteorological Office takes hourly records at some 14 sites and daily records at about as many others. These show substantial variations across the country, indicating that there are important local climatological effects. The annual figures also show quite large variation from year to year at any one site—sometimes the difference between successive years approaches 10%. Figure 6.4 shows some average figures for one site in Britain, expressed as the mean daily energy flux for each month. The quantity $T$ is the total energy flux and $D$ the diffuse component, for a horizontal surface; $S$ is the total energy flux for a vertical surface facing south. The best inclination for a receiving surface is often debated, but no single value can be given for this, for it varies with the location and the time of year. If the highest energy input is required in the winter months, a rule of thumb is that the inclination to the vertical should be a little greater than the maximum solar altitude at midwinter, giving a slope of, say $10°–15°$ greater than the latitude. Somewhat shallower slopes are desirable where the diffuse component of the radiation is predominant. When the highest energy input is required in summer, on the other hand, a horizontal surface may be near the optimum. This applies especially at higher latitudes, where the sun's position early and late in the day may be well to the north of the E-W line, and its direct radiation would not fall on a surface with a substantial inclination to the south.

Viewed together, then, the general characteristics of the solar radiation as an energy source are rather unattractive. The user has to contend with gross seasonal variations in the magnitude of the available energy, its duration during the day, and the direction from which it comes. These

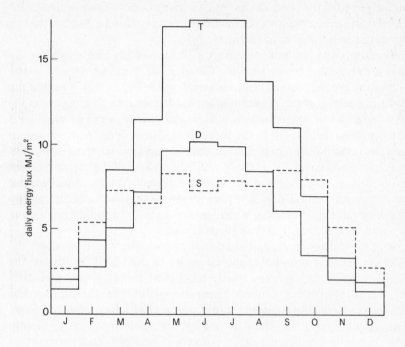

**Figure 6.4** Monthly mean daily energy flux, southern UK.

$T$ = total energy flux
$D$ = diffuse component for a horizontal surface
$S$ = total energy flux for a vertical surface facing south

factors are, however, predictable. Less easy to accommodate are the capricious fluctuations due to the weather. It is readily understandable that the only major uses of solar energy so far have been ones in which there is a measure of energy storage to act as a buffer between the supply and the demand.

### Heating by solar energy

Solar water heaters for domestic purposes are in widespread use in many parts of the world. The number in Japan is said to be several millions, and many more are operating in Australia and parts of the Middle East. The simplest kind of heater is illustrated in figure 6.5. Solar radiation is absorbed on a blackened surface, which becomes warmed and in turn heats water in contact with it. This heated water then passes, by natural convection or by being pumped, into a storage tank. Even as simple a system as this is engaged in quite complicated energy exchanges with the surroundings. As well as the solar radiation, the absorber surface receives long-wavelength radiation from the atmosphere and objects within its field of view. In turn, it emits long-wavelength radiation and loses heat by convection to the air in contact with it.

A simplified treatment of these processes will suffice to illustrate the behaviour. Suppose that the net rate of loss of energy per unit area from

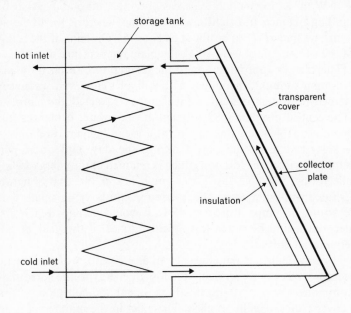

**Figure 6.5** Solar collector for domestic hot-water system.

the plate to the surroundings can be written as $U(T - T_s)$, where $T$ is the plate temperature, $T_s$ that of the surroundings, and $U$ is an overall heat-transfer coefficient. If the solar radiation flux is $P$ and the fraction of this absorbed by the plate is $\alpha$, the rate $P_e$ at which energy is removed from the system by the water in a steady state is given by

$$P_e = \alpha P - U(T - T_s) \qquad 6.1$$

per unit area. Thus the fraction of the incident solar energy which is transferred to the water is

$$\frac{P_e}{P} = \alpha - \frac{U}{P}(T - T_s). \qquad 6.2$$

Evidently, this fraction cannot exceed $\alpha$, and it has this value only when the plate temperature is the same as that of the surroundings; not a very useful condition. At other temperatures, the fraction diminishes with increasing $T$, and becomes zero when

$$T = T_s + \frac{\alpha P}{U}.$$   6.3

For a well-blackened plate exposed in still air, $\alpha$ might be about 0·9 and $U$ about 15 W/m$^2$ K, for moderate excess of temperature above that of the surroundings. Hence, in bright noon sunshine, where $P$ might be about 900 W/m$^2$, we find $\alpha P/U$ to be about 54 K ($= 54°C$). Thus, if the temperature of the surroundings were 20°C, the plate temperature would be over 70°C. (This is a fair approximation to the temperature reached by a black body in noon sunlight in summer with still air.) However, water cannot be obtained at this temperature, for if energy is extracted, the temperature falls. The central problem in the design of solar water heaters is thus to obtain a reasonable energy extraction at a useful temperature.

The solar radiation flux varies during the day, and an important characteristic will be the fraction which is retained of the energy falling on the collector over a whole day. For a simple heater, like that of figure 6.5, but operated without a cover, the fraction would be quite small. Even at a modest output temperature, say 45°C, it would perhaps reach 20% in summer, but would be much less at other times of the year at all but near-equatorial latitudes.

It will be evident that the efficiency in any given set of circumstances can be improved only by reducing the loss of energy to the surroundings. The commonest way of doing this is to shield the collector by one or more cover plates, usually of glass. These act in the manner of a greenhouse, admitting the short-wavelength radiation from the sun, to which they are nearly transparent, but impeding the long-wavelength radiation from the collector, to which they are nearly opaque. They also lower the rate of heat loss by convection. One glass cover reduces the net loss by about half in a typical case. There is a diminishing return from using more covers, and the advantages have to be weighed against the great increase in cost. It is found that one cover is more or less essential for heaters used in latitudes much beyond the tropics, but that two or more can rarely be justified except at latitudes approaching the polar circles, or where the highest efficiency is essential.

Another method of reducing the losses is by treating the absorber surface with radiatively-selective coatings. These give the surface a high absorptivity for solar radiation, but a low emissivity at their operating temperature. This is possible because of the widely-different wavelengths of the radiation in the two cases (the energy distributions with respect to wavelength are illustrated in figure 6.6). Selective coatings of various forms have been developed: thin films acting as interference filters and particulate suspensions or layers of semi-conductors which absorb at short wavelengths but do not much affect the emission characteristics of the substrate. Used with a single glass cover, coatings such as these may bring about a further halving of the losses. The resulting improvement in performance of the collector is substantial. Figure 6.7 shows some typical

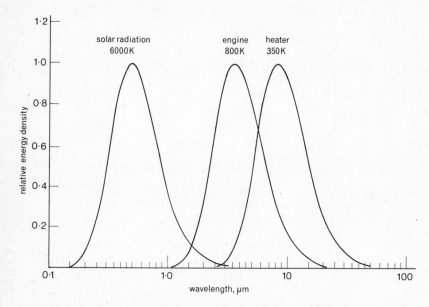

**Figure 6.6**  Approximate temperatures and energy distributions.

characteristics for operation in Britain. It may be noted that boiling becomes just possible in summer. Nevertheless, the output in winter is too low for the whole of a household's hot-water needs to be provided by collectors of reasonable size. It is usual to choose a collector size sufficient to meet all the requirements for several of the central months of the year, and at other times to boost the output temperature by more conventional means, the collector operating as a pre-heater. Thus the collector, though simple in itself, is usually part of a more complex system, containing in addition a storage unit and some auxiliary means of heating. Though each component has straightforward characteristics, the behaviour of the system as a whole is not at all simple, and its design calls for considerable engineering refinement if its performance is to be satisfactory in a wide range of conditions.

Examples of such systems, on a larger scale, are the arrangements for space heating provided in some 'solar houses'. A typical system is shown in outline in figure 6.8. The whole roof on one side of the house, and perhaps on a special extension, is now used as a collector, for the demands for space-heating in an ordinary dwelling are of the order of 500 MJ per day during the heating season at latitudes of 40°–50°. The collector communicates with a heat store in the basement, from which energy is drawn

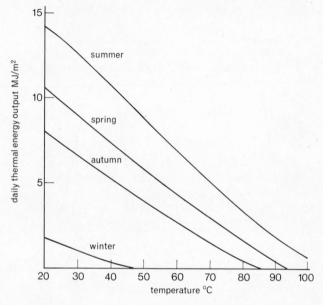

**Figure 6.7**  Seasonal variation in mean daily output of solar heater, southern UK.

and supplemented as required. The store is usually a chamber filled with water, or sometimes with boulders or crushed rock. A considerable volume is needed to provide the optimum amount of storage, particularly in cloudy environments, where there may be large fluctuations in the energy available from day to day, and sequences of consecutive days of low energy supply. Among methods under investigation for reducing the volume of storage units, is the use of media which undergo a change of phase (from the solid to the liquid state and vice versa) or a reversible chemical reaction at suitable temperatures. The energy changes involved in processes such as these are much greater per unit mass than in merely raising the temperature of an inert quantity of material. A simpler arrangement, promoted by Professor F. Trombe, Director of the Solar Energy Laboratory, CRNS, Odeillo, France, is the use of the 'solar wall', illustrated

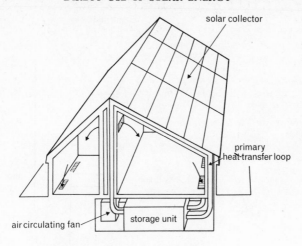

**Figure 6.8** Solar space-heating system.

in figure 6.9. Here the south-facing wall of the house is made unusually massive and serves both as a collector and a thermal store. Heat is convected directly into the house by allowing the air from within to circulate between the wall and its glass cover.

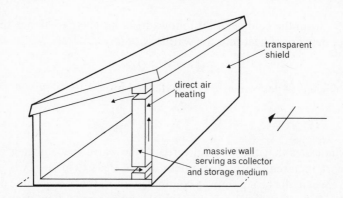

**Figure 6.9** 'Solar Wall' for space heating.

There has been an interest for some years in employing the heat pump for space-heating, and some quite successful installations have been described. The principle is illustrated in figure 6.10, showing that this device resembles a refrigerator in that it removes heat from a cold space (the surroundings) and discharges it into a warm space (the building). If work is required at a rate $W$ to drive the system and to extract heat at a

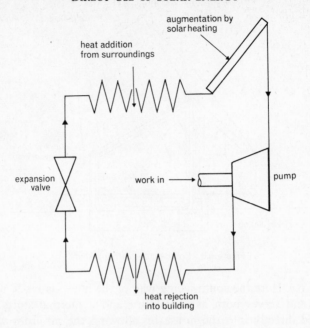

**Figure 6.10**  Solar-augmented heat pump.

rate $Q_1$ from the cold space, it follows from an energy balance that the rate of delivery of heat into the warm space is

$$Q_2 = W + Q_1.$$

A coefficient of performance C.P. is defined by

$$\text{C.P.} = \frac{Q_2}{W} = 1 + \frac{Q_1}{W} \qquad\qquad 6.4$$

This indicates that the heat delivered to the warm space is greater than the work required to drive the system, and to this extent energy is used to good effect in the heat pump. In common with all thermodynamic devices, however, the heat pump is subject to the fundamental limitations which are given formal expression in the Second Law. This relates the heat flows to the temperatures in the cold and warm spaces. In particular, it shows that the rate $Q_2$ cannot exceed the value $(T_2/T_1)Q_1$, where $T_1$ and $T_2$ are the absolute temperatures of the cold and warm spaces respectively. It follows that the highest possible value of C.P. is

$$(\text{C.P.})_{\text{max}} = \frac{T_2}{T_2 - T_1}. \qquad\qquad 6.5$$

Evidently this becomes greater, the smaller the temperature difference $(T_2 - T_1)$ between the warm and cold spaces. The only large sources of energy available for cold spaces are the atmosphere, the earth, and rivers and ponds. Heat exchangers must be used to extract the energy and, unless these are to be uneconomically large, there must be an appreciable temperature difference across them. Heat exchangers will also be required in the warm space, where again the energy can be delivered only if there is a temperature difference. Thus, the thermodynamic cycle is operating between a temperature lower than $T_1$ and higher than $T_2$, so that the high values of C.P. indicated by equation 6.5 cannot be realized. In a practical case, C.P. might be no more than about 3. Studies are in hand of ways of using solar collectors to raise the temperature of a cold space, perhaps a pond or a storage unit of the kind described earlier, so that a heat pump could then be used to deliver the energy collected at a temperature suitable for space heating, and operate with a respectable C.P. An increase in cold space temperature of 20 K (20°C) obtained by solar heating could enable C.P. values of about 5 to be obtained. Another means of augmenting the performance of a heat pump by using a solar collector is shown in figure 6.10.

Systems such as those described can be expected to come into more general use as fuel prices rise and resources dwindle. At high latitudes, however, an area the size of a roof cannot be expected to yield more than a part of the energy needs of the house unless steps have been taken to minimize those needs. The use of greater insulation, and the design of buildings so as to reduce heating requirements, are now being pursued much more vigorously than in the past. As the energy requirements become smaller, not only does the fraction which can economically be supplied by solar collectors rise, but the significance of the input of solar energy into the fabric of the house becomes greater. A factor which was formerly neglected, or at best acknowledged by rule-of-thumb, has now become a major element in building design.

Solar energy is being considered as a possible source in many other situations where heating at moderate temperatures is required. The provision of hot water for hospitals, schools, hotels, laundries, etc., may already be economic in most countries, and useful contributions might be made in numerous other situations, throughout community life, industry and commerce. The most promising applications are those in which the demand is matched with the variations in the supply of solar energy during the year, or, as in many of those noted, is at least roughly constant. In this respect, space heating is badly matched with the supply, for the greatest requirement comes at a time of year when the supply of energy is least. A better use, in some countries, would be in refrigeration and air-conditioning, which are most needed when the solar-energy supply is greatest. It has been found in parts of the United States, for example, that the maximum demand for electricity occurs in the summer, rather than in the winter, as at higher latitudes. This is traceable to the heavy demands for air-conditioning at this time. Absorption refrigerators, of a kind illustrated in figure 6.11, are basic elements in plant used for air-conditioning,

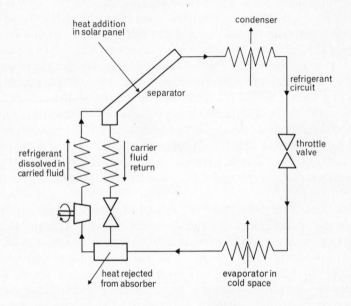

**Figure 6.11**     Solar-powered absorption refrigerator (SEM, figure 6.17).*

dehumidification and cooling of food stores. They have already been run with the heating of the separator provided by solar energy, but tend to be larger and less efficient than conventional plant, so that so far they have not been competitive. Another use, in which the demand is somewhat higher at times of highest energy supply, is the provision of fresh water. Solar stills, of a type represented in figure 6.12, provide an output equivalent to a few millimetres of rainfall per day over the area of the still. In these, water from a salt or polluted source is heated in a shallow tray under a transparent cover plate. Some of the vapour produced condenses in the cover, which is cooler than the tray because of its heat losses to the surroundings. The condensed water then runs down the cover into a collecting trough. Stills like this have been made and operated for years

* Brinkworth, B. J., 1972, *Solar Energy for Man*, Compton Press.

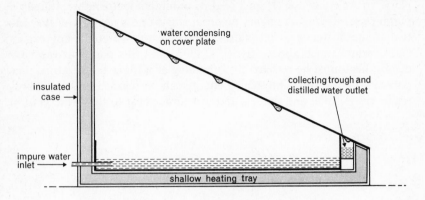

**Figure 6.12**  Simple solar still (SEM, figure 5.10).*

on a large scale, of the order of a hectare in area, and in places can provide potable water at costs competitive with conventional methods.

## Power from solar energy

It is a short step from the need for drinking water to the realization that water for irrigation is also needed most at times when the solar-energy supply is greatest. Solar-driven pumps have been made and operated, though again they are of low efficiency. This is mainly due to the fundamental limitations applied by the ubiquitous laws of thermodynamics. These set an upper limit on the efficiency with which energy in the form of heat can be converted into energy in the form of work. It is shown that no device for performing this conversion continuously can have an efficiency greater than the Carnot efficiency, given by

$$\eta_c = \frac{T_{max} - T_{min}}{T_{max}} \qquad 6.6$$

where $T_{max}$ and $T_{min}$ are respectively the maximum and minimum absolute temperatures of the working substance during its passage through the device. Consider, for example, the representative thermodynamic converter shown in figure 6.13. In this, a fluid (the working substance) is heated and turned into a vapour in the solar collector. It is at this point that its temperature is at a maximum. The vapour is then expanded through a turbine (or a reciprocating engine) to a low pressure. Before being returned to the heater again, it has to be cooled and condensed back to the liquid

* Brinkworth, B. J., 1972, *Solar Energy for Man*, Compton Press.

phase, in the condenser. Here it is at its minimum temperature. Finally a pump raises its pressure again before it starts on its next cycle. We saw that if the heater is a flat-plate solar collector, we could not expect temperatures much above, say, 60°C (333 K) at this point. At the condenser, heat must be rejected to something at a lower temperature. The only substantial sinks into which energy can be rejected are the atmosphere, rivers, lakes and the sea, and to transfer heat to these, the fluid in

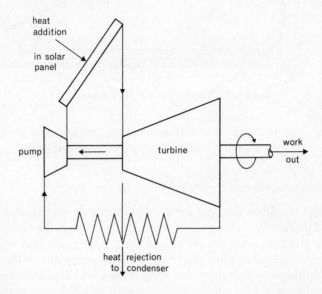

**Figure 6.13**   Elementary solar-powered heat engine.

the condenser would have to be at a temperature of not less than, say, 35°C (308 K). A perfect engine operating with these temperatures would, by equation 6.6, have an efficiency of about 0·075, or $7\frac{1}{2}\%$. Because of inevitable mechanical losses due to friction and non-ideal flow of the fluid, it would be difficult to get more than about half this efficiency in practice. When the efficiency of the collector is also taken into account, it is seen that, of the solar energy intercepted by it, a maximum of perhaps 2% or 3% might be converted into work. On a daily basis the energy recovery would be even lower. Moreover, this low performance is due principally to limitations of a fundamental kind, so that no amount of ingenuity in design could do anything to improve it significantly.

The only way of raising the efficiency is, by equation 6.6, to increase the maximum temperature in the cycle. This can be done by concentrating the energy collected over a large area onto a much smaller one. A well-known device for doing this is the paraboloidal reflector. The direct rays from the sun are brought by it to a focus. An object placed at the focus can be brought to a very high temperature, because its surface area, from which energy is re-radiated, is much smaller than the area of the reflector exposed to the sun. The largest reflector of this kind is used at the French solar-energy research station at Odeillo in the Pyrenees, for melting very refractory materials such as the oxides of zirconium, requiring temperatures in excess of 3500°C. Thus, such reflectors are capable, in principle,

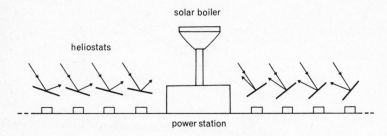

**Figure 6.14**  'Solar Tower' power plant.

of producing temperatures in an engine which would give very high efficiencies of conversion of heat into work. However, they are not without serious disadvantages. To obtain high temperatures they must have and retain good reflectivity and fidelity of shape. To be robust enough to withstand wind forces they must, if large, be quite massive. But the main difficulty is that they must be tracked round, quite accurately, to follow the apparent motion of the sun.

One way of overcoming this difficulty, as at Odeillo, is to keep the main reflector stationary and to reflect rays into it from a number of smaller reflectors (heliostats) which are themselves on moveable mountings. A refinement of this arrangement, first studied in the USSR, is the 'solar tower', shown in figure 6.14. Here the mountings of the heliostats provide the motion necessary to follow the sun, under the guidance of photoelectric tracking systems. In some schemes, a part of this motion is produced by the movement of the heliostats on carriages around a track surrounding the solar tower. Evidently there is a considerable price to pay for the complexity of these systems. Another solution is to use reflectors in the form of parabolic troughs, as shown in figure 6.15. The axis of the trough is orientated along a fixed east-west line and the only movement is in altitude. If the trough points towards the sun, the geometry is such that the rays are brought to a focus on the same focal line, whatever the time of day. If some degradation of performance is accepted, a fixed altitude may be used throughout the day, chosen so that the rays are intercepted best during the hours around noon. Only a minor adjustment of altitude need then be made, perhaps every few days, to allow for the sun's seasonal declination.

Reflectors such as these make it possible to operate engines and other converters with maximum temperatures which give them a respectable energy-conversion efficiency. Parabolic troughs should be straightforward

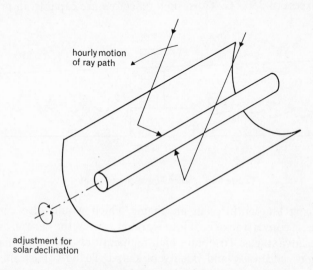

hourly motion
of ray path

adjustment for
solar declination

**Figure 6.15**　Parabolic trough collector.

to make in units of modest size (say 10 m long) and be replicated to make up a power plant of any required size. Proposals have been worked out in some detail for such plants, perhaps of a form represented in figure 6.16, to be built in desert areas. The energy collected at the foci of the troughs is transported away by vapour heat-pipes or a system of circulating liquid metals. Thermal storage units are necessary to accommodate variations in the energy supply during the day, and absence at night, and thus to match the supply with the demand. If plants such as these were built in remote areas, much of their value would be diminished by the cost of carrying the energy away to places where it is to be used, for example, by overhead transmission lines. It would be sensible to locate a purpose-built energy-consuming facility at a nearby point to minimize this cost. This would have the further advantage that there would be no overall movement of energy and no consequential environmental impact.

In consideration of the practicability of large-scale power generation from solar energy, the feebleness of the sun as a source becomes apparent. Focusing collectors cannot secure much of the diffuse component of the radiation, so that even in a favourable site, the supply may not be much greater than an annual total of about 5500 MJ/m$^2$. Of this, only a proportion, perhaps 1500 MJ/m$^2$ could be converted into work in a practicable system, though some of the remainder would be available for process heating. The output of a conventional power station of moderate size is

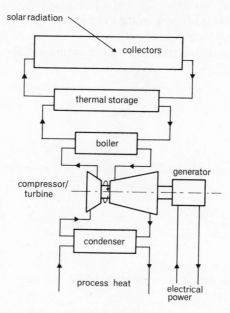

**Figure 6.16**   Elements of thermal-energy conversion system.

about $4 \times 10^{10}$ MJ per year, so that an equivalent solar plant would need an aggregate collector area of about $27 \times 10^6$ m$^2$, that is, about 5 km square. Though this may at first sight seem alarming, proponents of the idea point out that we are already accustomed to devoting much larger areas to energy production—in agriculture. To emphasize this comparison they have termed these power plants 'solar farms'. It has been suggested also that if they were built in desert areas which at present can serve no useful human purpose, it might be possible to cultivate the land in the shadow of the collectors and turn the area into a farm in the more conventional sense.

The construction of solar farms would be very demanding of materials at a time when concern is developing for the supply of some of those likely to be required. Many other possibilities have been explored for obtaining large collection areas at minimum cost. One such arrangement is that shown in figure 6.17. A low-lying uninhabited area (for example, the Qattara Depression in the Libyan desert near el Alamein) is connected by a channel or pipeline to the nearest sea, which may be considered to maintain a constant level. If the depression is in a desert area, there will be a very high rate of evaporation there (1700 mm per annum at Qattara). Thus when a steady state exists, there will be a difference in level between the open sea and the depression, and a continual inflow would be required to make up the evaporation. In the example cited this would be sufficient to drive a hydroelectric plant with a rated output of 4000 MW. Such a system is a solar-driven thermodynamic engine and its overall efficiency,

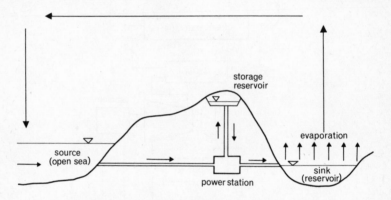

**Figure 6.17**   Helio-hydroelectric scheme, with pumped storage.

in terms of the energy falling on the depression, would be very low. Nevertheless, it provides a means of obtaining collection areas of very great size at lower cost than by any other means so far suggested.

One of the most convenient forms of energy is electricity, and the final stage of many thermodynamic conversion systems is an electrical generator. It is well known that it is possible to convert solar energy directly into electricity in devices having no moving parts. Some of these, such as the thermionic and thermoelectric generators, are found on inspection to be subject to the same limitations as more conventional engines. They involve a heating stage in which the radiant energy is first converted into heat, and a heat-rejection stage, required to keep some part of the system cooler than the rest. One kind of converter, in which there is no preliminary heating stage, is the photovoltaic cell, widely used in the powering of spacecraft. The operation of such a cell is illustrated in principle in figure 6.18.

If electrons in the cell material can be given energy in excess of a certain quantity $E_g$ (called the energy gap), they can be caused to leave the cell and to give up this energy in an external circuit before returning to the cell again. The required energy is obtained by the direct absorption of the solar radiation. An electron excited in this way would normally lose the energy immediately, reverting to its former state; the energy then appears as heat in the cell material. To be useful, the electron must be led away before this can occur, and in the photovoltaic cell this is brought about by an internal electric field introduced by doping the basic material with foreign atoms. One of the characteristics of radiation is that its energy is transported by its photons in packets or quanta, the size of which is

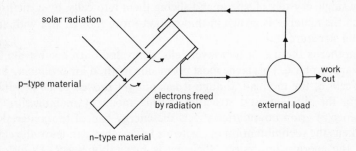

solar radiation

p–type material

electrons freed by radiation

n–type material

work out

external load

**Figure 6.18** Action of photovoltaic cell.

inversely proportional to the wavelength. Only the radiation whose wave-length is shorter than some critical value will have energy quanta greater than the gap energy $E_g$, so that the longer-wavelength radiation cannot be utilized by the cell. The energy of the shorter-wavelength radiation cannot be fully used either, any excess above $E_g$ being converted into heat. It follows that there is an optimum value for $E_g$ for any given spectral distribution of the energy, and a maximum efficiency with which the energy may be converted into electricity. For the solar energy the optimum value of $E_g$ is just over 1 eV* and the maximum conversion efficiency about 45%.

$E_g$ is a characteristic of the material of the cell, and several materials, mostly the intermetallic Group III–V elements, possess values close to 1 eV. Silicon is the most abundant of these and has been the basis of nearly all the photovoltaic cells used so far. To obtain the required characteristics, the material must be in single-crystal form, with doping atoms introduced at a low concentration—typically a few parts per million. The cells are then sliced from the crystal into wafers, and polished

* $1 eV = 1.6 \times 10^{-19} J$.

to a thickness of a few hundred micrometres. These requirements make the production of silicon cells extremely expensive, around £10 per watt of peak output at the present time. With this degree of refinement and at this cost, the overall energy-conversion efficiency is about 12%.

The provision of energy at such high cost is justifiable only where the cost of an alternative source is even greater. The only terrestrial purposes for which this is the case, so far have been the powering of active buoys and other remote navigational aids. Strenuous efforts are being made to bring down the cost of photovoltaic cells, but this would have to be reduced by a factor of about 100 to make them competitive with conventional converters. At least three approaches to cost reduction are promising.

The first seeks to avoid the complexity introduced by growing mono-lithic single crystals of silicon and slicing them into cells, by a method in which the material is grown in thin ribbon form, though still with single crystal structure.

Another is the use of thin-layer cells, deposited onto a substrate from the vapour phase. This is the most promising method for producing large-area cells at the moment, though it has not yet proved to be possible to obtain the single-crystal structure in this mode of manufacture. The presence of grain boundaries affects the energy gap of the material and quickens the recombination of electrons before they can leave the cell, so that the energy-conversion efficiency is inevitably lower. Diffusion of doping atoms along the boundaries also shortens the effective life. Silicon is not the most suitable material for this type of cell. Good results have been obtained with cuprous sulphide deposited in a layer less than 1 $\mu$m thick on a cadmium sulphide base, with reported conversion efficiencies up to about 8%; though in large-scale production, about half this value would be more probable. It is not yet clear whether the lower cost and possibly shorter life of cells of this type would compensate sufficiently for this lower efficiency.

A third approach is to accept that high-efficiency cells will be costly and to use concentrating reflectors so that the area of the cells is much less than the overall collection area. For this to be possible, cells must be able to operate at the very high energy flux at the focus of the collector. Again, it turns out that silicon is not the most suitable material, but high con-version efficiencies have been predicted (up to 20%) for cells consisting of a layer of gallium aluminium arsenide deposited on a base of gallium arsenide. With adequate cooling, cells of this type have operated with a concentration ratio of 2000, i.e. with a cell area only $\frac{1}{2000}$ of the collector area. It is too early in the development of these high-concentration cells to determine the optimum relationship between cell size and collector size for minimum cost, or what that cost would be.

One of the problems of using solar cells as a power source is the low voltage at which they produce maximum power, typically about $\frac{1}{2}$ V per cell. Many cells have to be operated in series to produce an acceptable voltage, and they need to be interconnected in parallel to avoid open circuits in the event of failure of one or more cells. Moreover, in many applications the output of the cells would not be directly matched to the demand, and their use would involve some element of energy storage. In those applications already existing, this has been provided by conventional storage batteries, which are bulky and expensive. Though proposals have been made for storage in the form of hydrogen (produced by electrolysis), and in large motor-driven flywheels, not enough system studies have yet been made to establish their cost and practicability.

The development of solar cells has reached its present stage largely through the requirement for operation of vehicles in space, without which it is very unlikely that a sufficient investment in them would ever have occurred. In this development, cost has not been a significant factor, and although this may have meant that the solutions could not be applied directly to mundane problems, the technology established through it forms an indispensable foundation for further work. This is being pressed ahead vigorously, for the prize would be of immense promise: a clean power source, providing its output in a very convenient form, with good conversion efficiency, having no moving parts, capable of being shaped to fit any surface, requiring no maintenance, and having a very long life. Further, it would be an intelligent solution, for it would be making use of the special characteristic of the sun as a radiative source—the unusually high energy of its photons.

## Photochemistry and photobiology

The energy carried by radiation is transported in units of finite size or quanta, and it is necessary to think of it as consisting of discrete particles, the photons. If the radiation is of wavelength $\mu\lambda$ m, the energy of one photon is about $1.24/\lambda$ electron-volts ($1\,\text{eV} = 1.6 \times 10^{-19}$ J). In the photovoltaic cell, this energy is communicated to an electron on absorption of the photon. Other absorption processes may have different consequences, such as the lysis of molecules, perhaps leading to recombination in more desirable forms. Photochemical processes with high yields are already widely used for the manufacture of products such as solvents, insecticides, pharmaceutical chemicals, and vitamins. Few of these processes could proceed at significant rates in sunlight, however, for the photon energies are in general still too low, so that ultraviolet sources are employed in practice. What is needed for the utilization of solar energy through photochemistry, is the recognition of some process in which photons with energies throughout the solar spectrum can be absorbed and lead to stable species from which the energy may be recovered at will. Unquestionably, the most desirable would be one leading to the photolysis of water into hydrogen and oxygen, thus yielding a storable and transportable fuel from a cheap and abundant raw material. The process might be represented by equation 6.7.

$$2\,H_2O + \text{radiation} \rightarrow 2\,H_2 + O_2 \qquad\qquad 6.7$$

Electrolysis shows that about 3 eV of energy is required to split a water molecule. This could be achieved, in principle, by a photon with wavelength less than about 0·4 $\mu$m. However, water is virtually transparent at these wavelengths, that is, there are no mechanisms by which a water molecule can absorb photons in this region. This is typical of the frustration, if not the irony, which surrounds the subject. Every promising energy-storing reaction so far studied is subject to some effect which makes it unsuitable for practical use. Sometimes the products are highly reactive and re-combine immediately; sometimes they also absorb radiation and are reduced to less useful products; where a suitable reaction will proceed, it does so at a very low rate or in a region of the spectrum in which the solar radiation is deficient in energy.

Though striking advances have been made in photochemistry in recent years, much remains to be understood about it. Long development has enabled one photochemical process—photography—to be sensitized by dyes so that it may proceed at wavelengths right out into the infra-red. These sensitizers are evidently capable of retaining the absorbed energy long enough for more than one photon to take part in the process, probably by the occurrence of relatively long-lived triplet states. The search continues for sensitizers which would store energy over a large part of the solar spectrum and mediate the lysis of water or some other agent, with a good yield of stable components which can be recombined at will. To have long-term value, the sensitizer should be a mere catalyst, being re-generated in the process also. The required reaction would occur in two steps: a photochemical reduction

$$2\,H_2O + 4M + \text{radiation} \rightarrow 4H^+ + 4M^- + O_2, \qquad 6.8$$

followed by an ordinary chemical oxidation,

$$4H^+ + 4M^- \rightarrow 4M + 2H_2 \qquad\qquad 6.9$$

in which the catalyst M is also reconstituted. There have been promising leads, but so far no catalyst has been found which operates with acceptable efficiency.

It is probable that a practicable solution, if one is ever developed, will be via a photochemical cell, illustrated in principle in figure 6.19. Photons are now absorbed on an electrode, comprising a semiconductor having a band gap enabling a useful fraction of the solar energy to be acquired by electrons. Hydrolysis then occurs at the electrode surface:

$$2\,H_2O + radiation \rightarrow 4\,e^- + 4\,H^+ + O_2 \qquad\qquad 6.10$$

The hydrogen is released at the dark electrode by electrons returning via the external circuit:

$$4\,e^- + 4\,H^+ \rightarrow 2H_2 \qquad\qquad 6.11$$

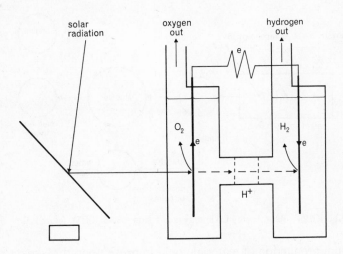

**Figure 6.19**   Action of photochemical cell.

This kind of process has been demonstrated, but with very low efficiency.

   Meanwhile, natural photosynthesis offers an enviable example of the possibilities in this area. The essential overall reaction there is representable by the equation

$$n\,CO_2 + n\,H_2O + radiation \rightarrow n\,CH_2O + n\,O_2, \qquad\qquad 6.12$$

though the carbohydrate molecule is used as a building block for the multitude of more complex organic substances which forms the plant. It has been shown that none of the oxygen liberated comes from the $CO_2$ inspired by the plant, so the basis of the reaction is again the lysis of water. Since the process can be driven by photons with energies at the red (low-energy) end of the spectrum, it appears that several (probably eight) photons are used per carbohydrate molecule synthesized. This would lead to an overall efficiency of solar-energy conversion of about 10%, which can be obtained under laboratory conditions.

Energy can be recovered from plant materials by direct combustion, or by fermentation and other processes, leading to combustible liquids such as alcohols, or gases such as methane. These processes can yield about 20 MJ of energy per kilogram of plant material. Vigorous plants growing under tropical conditions produce material at a rate which would permit about 3% of the incident solar energy to be recovered as heat on an annual basis. It remains to be seen whether the breeding of plants specifically for

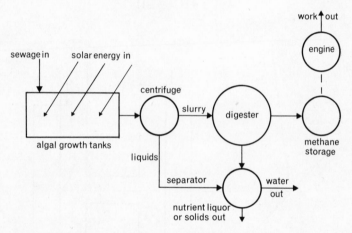

**Figure 6.20** Photobiological energy conversion system (SEM, figure 9.5).*

maximum production of material could improve upon this significantly. Some recent studies of the growth of plants for fuel have called attention to the small proportion of the land area which would be required for this, even in developed countries. However, in these studies there has been an almost total neglect of the problems of engineering a power-producing system on this basis. Much closer study is needed of factors which are disadvantageous: the energy used within the system, for felling, transporting, drying and processing this very low-grade fuel; the efficiency of energy conversion into work; and the environmental consequences of discharging the products of combustion or fermentation.

Another possible way of using natural photosynthesis would be via the lower plants, as in figure 6.20. Unicellular algae are grown in tanks exposed to solar energy and nourished on waste products such as sewage. In the growth tanks there develops a symbiosis between algae and bacteria,

* Brinkworth, B. J., 1972, *Solar Energy for Man*, Compton Press.

mutually exchanging oxygen and carbon dioxide. The material produced is continually taken off and centrifuged, yielding a largely solid part which is passed to a closed digester where there are anaerobic bacteria capable of converting it into methane. The liquid part is separated into usable water and a nutrient liquor which might be used as a fertilizer. Though the system may seem elaborate, it uses established technology, and is in reality only a concentration of natural processes under close control. The feasibility has been demonstrated at pilot scale, but the energy-conversion efficiency, based on the area of plant exposed to radiation, is again unattractively low.

## The prospects for solar energy

In the minds of many people, solar energy is seen as a possible replacement, indefinitely renewable, for the conventional sources of power—mainly the fossil fuels—or as an alternative to nuclear power. It has not yet been demonstrated, however, that these substitutions are credible. This is not because of overwhelming technical difficulties; many of the systems whose principles have been outlined here have been thoroughly studied and even operated in prototype form. Though there would be a need for some further research and much development before they could be put into commission, no insuperable problems of a technical nature can be discerned which would rule out solar power in principle. Indeed, if it were essential that the world be powered in this way, we can say that we know how to do it, which few would claim to be the case at the present time for other long-term options, such as fusion power. The main obstacle is one of scale. None of the systems described is capable of converting more than about 20% of the energy arriving at the earth's surface into mechanical or electrical work, and this means that large areas would be needed for collectors. There are uninhabited regions amply large enough for these, but in most cases they are far from the main centres of consumption, so that a big investment in transmission equipment would be needed. Large collectors would have a large capital cost, almost certainly greater than the capital cost of conventional power plant of the same capacity. However, most solar systems are relatively straightforward from an engineering viewpoint and would be expected to have very long service lives, so that in the long term the cost of power from them might be less. A more serious restriction might arise in the necessary provision of raw materials for construction, which would be required in much greater quantity for a solar power plant than for conventional plant of the same rating. Proponents of the 'energy accounting' approach to system selection have been very critical of nuclear power plant because of the energy used in fabricating it and processing its fuel. It might be that large-scale solar-energy plant would be no better in this respect because of its size.

The use of ponds and reservoirs as solar-energy collectors of lower cost is offset by the even greater size required, because of their inherently low operating temperature and consequent low thermodynamic efficiency. On the other hand, they do embody the element of energy storage which would be crucial for the satisfactory operation of a system in which the supply cannot be adjusted to match the demand. This is a feature also of photobiological systems and some photochemical ones, in which the collection, conversion and storage functions are combined.

The case for centralized power generation, and distribution to consumers is strongest in the more highly-developed countries, whose development is based on urbanization and industrialization. Many of these also have the highest population densities and can least spare any space for large-scale collectors. For them, the most desirable options are the ones offering the highest efficiency of energy conversion, and hence the smallest collection area for a given output. As seen at the present, these are thermodynamic converters with concentrating collectors and photo-voltaic cells. So far, the second is altogether too expensive, and the problems of storage for an electrical output are a further handicap. The development of cheap polycrystalline cells of large area, and of means of storing hydrogen produced by electrolysis of water on a large scale, are perhaps the most promising routes for photovoltaic generation. In suitable terrain, pumped storage reservoirs may provide the cheapest means of energy storage for thermodynamic systems. Direct thermal storage, though inconvenient, is a straightforward option combining simplicity with advantages of scale (losses to the surroundings, being roughly proportional to surface area, are accordingly proportional only to the $\frac{2}{3}$ power of the volume). The indications are that the long-term cost of power from a large solar-thermodynamic station, though greater than that of conventional plant at the present time, would become competitive following only a modest further increase in the cost of fuels. Such systems must, therefore, be taken seriously. For countries lying beyond the tropics, however, there would be difficulties in integrating them with other power sources in a national system, because of the seasonal variation of output.

In recent years there has been much concern about the effects of large-scale technological operations on the environment. It should be understood that solar-power plants would not be entirely without environmental effects. Operation of collectors covering a large area would alter the energy balance of the region relative to its former value. Solar radiation is partly reflected from the terrain and partly absorbed. The reflected part normally has a spectral distribution not grossly different from that of the incident radiation, and in its return through the atmosphere it is absorbed only weakly in the immediate vicinity of the site. The absorbed radiation, on the other hand, warms the surface, so that this part of the energy is transmitted by convection and long-wavelength radiation into the air close to the site. In designing the plant, it might be possible to ensure that the net energy input into the surroundings was similar in magnitude to its former value in the undeveloped site, though its nature would be somewhat altered, to the extent that the collectors were operating at different temperature from that of the former ground surface. The input into the ground itself would almost certainly be different also.

If the plant is a thermodynamic one, the major part of the energy collected would have to be rejected eventually into the surroundings. Even if this energy were used first for process heating, it must finally be discharged above ambient temperature. If it is discharged into a river, this may produce 'thermal pollution' in the same way as the outfall of any other power station. This is probably the most important single environmental effect, though discharge into the atmosphere via cooling towers would also alter the former energy balance of the region to some degree. It is difficult to say whether the sum of these effects would be significant in a particular case, for they would be unusual in being operative over a very large area. Experience with schemes of a comparable scale is perhaps to be found only in afforestation and certain strip-mining operations.

At the consumers' end, the remaining part of the energy will also find its way into the environment. Thus, the operation of solar-power plants would release additional energy into the biosphere, and in this respect would be no different from using any other power system. It would, however, not release any carbon dioxide or other products of combustion, and this might prove to be an overwhelming advantage.

These and other considerations have led some to the view that the proper scale of solar-energy utilization should be small and local, rather than large and central. The energy is already distributed and it might be sensible to use it where it falls. There is certainly a strong case for obtaining a solar contribution to domestic water-heating, air-conditioning and space-heating, and studies are being made of possible ways of meeting other domestic energy requirements, at least partly, in this way. The achievement of the 'autonomous house', entirely self-sufficient in energy, may still be some way off, but intermediate solutions may be within reach soon. Moderate-scale systems for schools, hospitals, factories and farms are now receiving attention.

It is probable that the best opportunities lie in the developing countries. Often well endowed with solar energy, these have been badly hit by rising fuel costs and desperately need power for the provision of work and amenity among scattered populations with poor communications. Solar-powered devices could substantially raise productivity in such local enterprises as weaving, timber sawing, paper and board milling, food-preserving, fresh-water supply and a host of others. Through these it may be possible to revitalize the village and stem the disastrous drift to the town. All the hot-water needs of a clinic and the air-conditioning of its wards and theatres could be provided by the energy which falls on its roofs. The roof of a dwelling could furnish power for a loom or a lathe, that of a barn enough for a pump, able to irrigate 10 hectares of land. A school's radio and TV receivers could be powered by a device the size of a blackboard; a solar still the size of a duck-pond could yield drinking water for a village. These possibilities call for responses from the more developed nations, extending across the spectrum from technical-aid programmes on the one hand to new export business on the other.

There are encouraging signs that the research and development effort necessary to clarify the possibilities for solar-energy use will be forthcoming. A US National Science Foundation report of 1972 recommended a commitment of \$3500 million in this area, with the contention that solar energy could provide 35% of the nation's heating and cooling requirements for buildings, 40% of its fuels, and 20% of its electricity. Work is proceeding under the programme of research allied to national needs (RANN), with funding having reached about \$50 million per year. A comparable programme has been started in Japan (Operation Sunshine) and a considerable effort is being applied in Australia and the Soviet Union. There are reports of mounting activity throughout the world. It is noteworthy that together these programmes represent an investment in solar energy comparable with the cost of the Apollo moon-landing programme. As the world awakes to the need to husband its resources and to preserve a healthy environment, this will seem a wise and timely investment.

## FURTHER READING

For a more comprehensive treatment of the principles of solar-energy utilization, see

B. J. Brinkworth (1972), *Solar Energy for Man*, Compton Press.

Two useful earlier works are:

F. Daniels (1964), *Direct Use of the Sun's Energy*, Yale University Press.
A. M. Zarem and D. D. Erway (1963), *Introduction to the Utilization of Solar Energy*, McGraw-Hill.

Articles and research papers are to be found throughout the periodical literature of science and engineering. Two specialist journals dealing exclusively with solar energy are *Solar Energy* (Pergamon) and *Applied Solar Energy* (Allerton Press, N.Y.; this is a cover-to-cover translation of the Russian journal *Geliotekhnika*). Much of the original work from all parts of the world is brought together at international conferences, and the proceedings are a convenient reference source. Some of the most comprehensive are those of two UN conferences:

*New Sources of Energy*, Rome, 1961;
*The Sun in the Service of Mankind*, Paris, 1973.

# Index